TAXONOMIC ANALYSIS IN BIOLOGY

Taxonomic Analysis in Biology

Computers, Models, and Databases

**Lois A. Abbott, Frank A. Bisby,
and David J. Rogers**

New York   **Columbia University Press**   1985

Columbia University Press
New York   Guildford, Surrey
Copyright © 1985 Columbia University Press
All rights reserved

Printed in the United States of America

Library of Congress Cataloging in Publication Data

Abbott, Lois A.
  Taxonomic analysis in biology.

  Bibliography: p.
  Includes index.
  1. Biology—Classification—Data processing.
I. Bisby, F. A.   II. Rogers, David J. (David James),
III. Title.
QH83.A197 1985   574′.012   85-4188
ISBN 0-231-04926-9
ISBN 0-231-04927-7 (pbk.)

Contents

Preface   **xiii**

Introduction   **1**

    Chapter Contents, 3; Approaches to Taxonomy—Our Philosophy, 6; Taxonomy and Computers, 7

Part I   Standard Taxonomy   **9**

Chapter 1.   Taxonomy: An Information System for Biology   **11**

    The Taxonomic Information System, 11; Taxonomy in Today's World, 13; Suggested Readings, 19

Chapter 2.   Taxonomic Products   **20**

    The Classification, 21; The Descriptions, 23; The Nomenclature, 27; The Identification Aids, 30; Suggested Readings, 32

Chapter 3.   Standard Taxonomic Analysis   **34**

    Introduction: Objectives and Literature Survey, 36; Materials and Methods: Specimens, Data and Taxonomic Characters, 40; Results: The Classification, 47; Conclusions: The Publications, 52; Suggested Readings, 54

Part II   Theoretical Bases for Taxonomic Work   **57**

Chapter 4.   Structure of Taxonomic Data   **59**

    Tabular Data Structures, 59; Data Types, 63; Formulating Descriptor States, 67; Nontabular Data Structures, 69

Chapter 5.   Models and Algorithms   **72**

    Models for Taxonomic Hierarchy, 74; Algorithms for Producing Taxonomic Hierarchies, 112; Suggested Readings, 117

Part III   Computer-Assisted Taxonomic Analysis   **119**

Chapter 6. Character Analysis   **121**

Classificatory Value, 121; Types of Characters, 122;
Structural Properties of Classificatory Characters, 124;
Correlations Among Classificatory Characters, 129;
Character Selection and Weighting, 140; Further Devel-
opment of Character Analysis, 142; Suggested Readings,
142

Chapter 7. Phenetic Classification   **144**

Agglomerative Two-Step Cluster Analysis, 145; Optimization Cluster Analysis, 165; Ordination, 166; Canonical
Variates Analysis (MANOVA), 171; Choice of Methods, 178; Suggested Readings, 187

Chapter 8. Diagrams of Variation Pattern   **188**

Scatter Diagrams, 188; Multivariate Data Tables, 190;
Physical Models, 193; Dendrograms, 197; Linkage Dia-
grams, 202; Minimum Spanning Trees, 207; Combined
Tactics, 208; Suggested Reading, 210

Chapter 9. Identification   **211**

Advances in Field Guides, 212; Computer-Produced
Identification Aids, 213; Identification by Computer, 222; Conclusions, 225; Suggested Readings, 226

Chapter 10. Phylogeny and Cladistics   **227**

Phenetics and Phylogeny, 229; Narrative Methods, 233;
Parsimony, 237; Cladism, 244; Compatibility, 251; Suggested Readings, 254

Part IV   Computer-Assisted Database Management   **255**

Chapter 11. Database Management   **257**

Computing Facilities, 257; Databases, 259; Computerized Database Management, 261; Choice of Database
Management Systems, 281; Suggested Readings, 282

Chapter 12. Taxonomic Databases   **283**

Types of Databases, 283; Examples of Databases, 293;
Suggested Readings, 313

Chapter 13. Goals for the Future   **314**

Gaps in the Computer-Based Information System, 314;
Two Systems or One?, 318; The Future: Brass Knobs
and Pure Speculation, 319

References   **321**

Index   **333**

List of Figures

1.1 The retail and diffusion models of information services
     provided by taxonomists.   18
2.1 Portion of a technical description of the Maple family.   25
2.2 Part of a popular description of the Maple family.   26
3.1 Leaf forms used to determine characters in some Legumes.   43
4.1 Summary of data types.   65
4.2 Example of data types in a botanical study.   66
4.3 Histogram and frequency polygon used to establish
     descriptor state ranges.   69
4.4 Diagram of pairwise, hybridization relations among
     species of *Geum*.   70
5.1 Euclidean distance in two-dimensional space.   77
5.2 Calculation of Euclidean distance between two species.   77
5.3 Angular distance measure in two-dimensional space.   79
5.4 Taxonomic versus statistical correlation.   82
5.5 Rotation of axes in PCA to secure line of best fit.   84
5.6 Standardized values and loadings on components from a
     PCA on standard blood variables.   86
5.7 Diagrams of objects grouped according to their
     transformed scores following PCA.   88
5.8 A graph (undirected).   94
5.9 Minimum spanning tree for five objects.   98
5.10 A dendrogram and a directed tree.   99
5.11 Calculation of maximal information when all states
     have equal probability.   102
5.12 Information calculations using two different distributions
     of character states.   103
5.13 Calculations for character comparisons using information
     measures.   105

5.14 Information measures of group diversity. 110

5.15 Calculation of changes in information. 112

6.1 A histogram demonstrating bimodal distribution of character states. 127

6.2 A histogram showing the unimodal distribution of pod length in a vetch species. 128

6.3 A bar diagram showing distribution of samples from 37 species for a metric morphological character. 130

6.4 Table showing how correlated characters define three subfamilies of the Leguminosae. 131

6.5 A Venn diagram representing information held exclusively or shared by two characters. 133

6.6 An example of high linear correlation between two characters. 137

6.7 Diagram illustrating taxonomically useful correlation. 138

6.8 Diagram of characters separated on the basis of three characters. 139

7.1 Diagram of overall procedure in agglomerative two-step cluster analysis. 146

7.2 Example of a graph containing a sub-graph or linkage group. 152

7.3 Diagram of the linking of two items with dendogram showing the two items clustered 152

7.4 Formation of a three-member cluster. 153

7.5 Linking an item with a previously formed cluster. 153

7.6 Linking two previously formed clusters. 153

7.7 Formation of an internal link. 153

7.8 Dendogram of a long, thin cluster compared with a more dense, compact cluster. 154

7.9 A linkage diagram for the analysis of relationships among species of *Crotalaria*. 156

7.10 The fusion of an item with a cluster based on average distance and the fusion of two clusters. 158

7.11 McNeill's dendrograms resulting from six different forms of cluster analysis applied to the same data. 160

7.12 The fusion of an item with a cluster using a central point algorithm and the fusion of two clusters. 162

7.13 Means for finding the central point in the median cluster analysis method. 162

7.14 Means for finding the central point in the true centroid cluster analysis method. 163

7.15 Diagrams showing means by which reversals can occur in central point clustering methods. 163

7.16 Groupings within the order Ericales produced by true centroid cluster analysis. 164

7.17 Results of PCO used to analyze morphological diversity
     in populations of grass snakes.   168
7.18 Results of MDS used to analyze variation between two
     species of *Silene*.   172
7.19 Determination of the axis that maximizes discrimination
     between two clusters using Discriminant Function.   174
7.20 Demonstration of the effect of correlations within clusters
     on canonical distances between clusters in MANOVA.   175
7.21 Drawings of the skull of a shrew, a map showing distribution
     of the shrew population, and results of canonical variates
     analysis showing groupings of the various island populations
     on the basis of the skull measurements.   176
7.22 Various cluster shapes.   178
7.23 Example illustrating space dilation that occurs with central point
     and group-average clustering methods.   179
7.24 Example illustrating the space contraction effect of chains
     in single-linkage clustering analysis.   179
7.25 Determination of the α gap between clusters in a dendrogram.   180
7.26 Illustration of chaining versus balanced heirarchies.   181
7.27 Illustration of the effect of small changes in distances on
     single-linkage and group-average dendrograms.   182
7.28 Example of consensus trees.   184
8.1 A two-dimensional scatter diagram illustrating separation of
     two distinct groups of plants on the basis of two characters.   189
8.2 An illustration of a scatter diagram drawn by a computer
     graphics plotting program.   191
8.3 Scatter diagram showing trends in gastropod shell dimensions.   192
8.4 A hypothetical cause that illustrates clustering based on three
     principal axes by plotting, in succession, their three possible
     pairwise combinations.   194
8.5 A solid model that represents positions of objects
     in three dimensions.   195
8.6 A pictorialized scatter diagram.   196
8.7 A perspective view of a three dimensional plot drawn
     by a computer.   198
8.8 Stereoscopic pairs generated by a computer to assist in the
     analysis of points plotted in three dimensions.   199
8.9 An explanatory diagram indicating the components
     of a dendrogram.   200
8.10 A truncated dendrogram.   200
8.11 Two identical dendrograms.   201
8.12 Steps in drawing a series of linkage groups.   203
8.13 Criteria for drawing circles to indicate closely linked objects.   205

8.14 Three different linkage diagrams any of which could be
depicted by the nodes in the dendrogram.   206
8.15 Comparisons of representations for simple two-dimensional
data using a scatter diagram, minimum spanning trees, and a
linkage diagram.   208
8.16 An example of a minimum spanning tree combined with
a scatter diagram.   209
9.1 Diagrams of balanced and unbalanced keys.   215
9.2 Example of a condensed format for a key   216
9.3 Example of an edge-punched card key.   219
9.4 A punched card key for the genus *Astragalus*.   220
9.5 Example of the printout from a computer polyclave
identification program.   223
9.6 Diagram of taxon boundary relationships in an identification
by a matching method.   224
10.1 The Salmon, Lungfish, and Cow controversy   228
10.2 A diagram illustrating that phylogenetic pattern adds the
time dimension to phenetic groupings.   230
10.3 Two very different phylogenies produced from a single
cladogram by varying the timing.   231
10.4 A cladogram that corresponds to any of several trees.   232
10.5 Several trees that are possible when A, B, and C represent
fossils.   232
10.6 The relationship between time and phenetic distance if, and
only if, the evolutionary rate is constant.   233
10.7 A balloon diagram illustrating relationships among
Magnoliatae.   234
10.8 A diagram of narrative relationships in the ferns.   235
10.9 A line diagram showing relationships among genera
of Tamaricaceae.   236
10.10 A narrative tree-like structure.   236
10.11 A scheme for depicting present-day groupings of taxa as
a cross-section through a phylogenetic tree.   238
10.12 An undirected tree and a directed tree.   239
10.13 The Manhattan measure of evolutionary distances.   240
10.14 The use of Steiner points in producing the shortest
minimum spanning tree.   241
10.15 The effect of using different starting points when
producing evolutionary trees from a Wagner network.   241
10.16 Three possible evolutionary trees derived by the Wagner
tree method applied to 23 genera.   244
10.17 Venn diagrams and corresponding cladograms.   246

10.18 An example of conflicting synapomorphies. 247

10.19 Resolving a conflict by majority vote. 248

10.20 A classification based on a cladogram. 249

10.21 Possible combinations into undirected trees of two binary characters. 252

10.22 An undirected tree formed by the largest clique of 14 compatible characters. 253

11.1 Two examples of forms for data collection. 262

11.2 Flow chart of a data storage and retrieval system. 265

11.3 Sample table from a database for *Crotalaria*. 267

11.4 Card image output resulting from item definition in TAXIR. 268

11.5 Examples of query outputs in TAXIR. 272

11.6 Examples of a query in a hierarchic database, GIS. 276

11.7 Sample relational database tables. 279

12.1 A distribution list from a biogeographic database. 285

12.2 Computer generated species distribution map. 286

12.3 Lists from a bibliographic database. 287

12.4 Illustration of character dependencies. 290

12.5 Examples of coded character descriptions. 292

12.6 DELTA coded form of a genus description and natural language description reproduced by CONFOR. 294

12.7 Part of a catalogue produced by TAXIR. 297

12.8 Specimen data input for museum collection and computer produced labels for the specimens. 299

12.9 Example of a SELGEM query. 300

12.10 Coding of descriptors for Colombia National Herbarium. 302

12.11 Sample printout from Colombia National Herbarium. 303

12.12 Examples of MDS record cards. 304

12.13 Flowchart showing information services provided by the Vicieae Database Project. 311

13.1 An idealized scheme for the use of descriptive information in a variety of taxonomic activities. 315

List of Tables

4.1 Tabular Data Structure for P Items and n Descriptors.   61

4.2 An Example of Ordered Data.   64

6.1 The Characters Analysed, Listed in Descending Order of Information Contribution.   125

6.2 An Example of Two Characters Having Nested Information.   135

7.1 A Contingency Table.   148

Preface

Several premises underlie the ideas expressed in this book. First is the realization that taxonomy is still a much needed science. It has its roots in the oldest of human endeavors in what we now call the biological sciences, but it is still unfinished. Further, it functions as a service to all biologists, because its purpose is to organize the information about living organisms, which is basic to the many special fields of biology.

Second, we believe that taxonomic work can be greatly assisted by modern computer methods. But, we would add, maximum benefits can be obtained from computers applied to taxonomy if, and only if, the new methods are adjuncts to, and not substitutes for, the practical methods that working taxonomists have developed over many centuries. Third, we believe that computerization itself has led to important new understandings of the theory underlying the taxonomic process.

The impact of computers on the practice of taxonomy has been somewhat different in the two English-speaking countries that the authors represent, the United States and the United Kingdom. In the United States most biologists think of 'computerized taxonomy' as the same as 'numerical taxonomy,' a name introduced by Sokal and Sneath in their germinal book first published in 1963. The term 'numerical taxonomy' was originally applied to a group of methods based, largely, on multivariate statistics, as exemplified by the set of computer programs known as NT-SYS. The emphasis is on quantitative measurement data applied without weighting, the intention being to maximize objectivity.

Another group of methods developed in the U.S. and sometimes referred to as 'taximetrics' (Rogers 1963), was derived pragmatically from analysis of the thought processes that taxonomists actually use to arrive at classifications, keys, and other products—including means of analysis to determine taxonomically useful characters and methods for information storage and retrieval. The theoretical and mathematical bases for these methods come from various aspects of set and graph theory. Development of these methods occurred at the New York Botanic Garden, Colorado State University (Ft. Collins), and the University of Colorado at Boulder, where one of us (DJR) was the principal investigator. A team effort was employed throughout, and members of the team included mathematicians Taffee Tanimoto and George Estabrook, programmer Bob Brill, and taxonomists Henry Fleming and Michael Wirth. The other authors were at one time or another associated with the Boulder group.

Some methods, developed chiefly in the United Kingdom, devote more attention to the use of qualitative data. Criteria based on qualitative characters have always been important in the classifications produced by standard taxonomic methods; it was necessary to devise computer-assisted methods for taxonomy that would make use of them. For example, the Williams and Lance group, once at University of Southampton, England and now in Canberra, Australia, has developed a number of methods and programs based on information theory rather than statistics. These allow for efficient use of qualitative as well as quantitative data.

In addition to computer-assisted analysis of taxonomic data, the new methods now available for information management have led to a resurgence of interest in developing and utilizing large taxonomic databases. Although the inability to obtain funding for the Flora North America project dealt a major blow to plans in the United States, major efforts are underway in Europe and other places (Allkin and Bisby 1984) to create large-scale, computerized taxonomic databases.

The difficulties of doing a book with three authors—especially when they live in three different places separated by thousands of miles— are obvious. The process took a long time and had its frustrations, but also its rewards; we now have a deeper understanding of just what the taxonomic process entails. This is due in part to contributions from others and we are grateful for this help. We want to thank especially the following for reading preliminary drafts of some chapters: P. Bryant, G. Estabrook, C. J. Humphries, N. Maxted, and J. Cresswell. Also those who provided illustrative materials for the chapter on databases including the Office of Computer Services at the Smithsonian Institution in Washington D.C., the Museum Documentation Association in Duxford, England, A. Gomez-Pompa and his group at Vera Cruz in Mexico, E. Forero at the Colombia National Herbarium in Bogota, R. S. Hoffmann at the University of Kansas Museum in Lawrence, W. A. Weber at the herbarium at the University of Colorado in Boulder, and T. J. Crovello at Notre Dame, Indiana. We much appreciate the work of Ryland Loos, illustrator for the Biological Sciences Department at State

University of New York at Albany, who drafted the many figures and of
Robert Speck, also of SUNY-Albany, who did photography for the figure
preparation.

L.A.A.
F.A.B.
D.J.R.

Introduction

This book is aimed both at taxonomists and at people in related fields—paleontologists, evolutionary biologists, population geneticists, biogeographers, phytosociologists, plant and animal breeders, ecologists, and others studying natural history. It is important that people in those related fields, who must work with classifications, understand the assumptions and methods of analysis on which the classifications are based. Without these they can neither use the information in a sophisticated manner nor justify the interpretations made of it.

Our subject is taxonomic analysis in biology, that is, how biological taxonomy is done: how it has been done, using methods developed over the past several centuries; and how these methods are now being affected by the advent of computers. According to one of the editions of Webster's dictionary taxonomy is "the science of classification; the laws and principles covering the classifying of objects." It is this that forms the major focus of our book.

We shall use the word taxonomy to mean the *process* of classification. However, it is not infrequent to see the term taxonomy used to mean the *result,* the classified information pertaining to a group. For example, there are courses and books entitled "Taxonomy of Flowering Plants" that provide facts: the names of major plant families (legumes, composites, roses, orchids and so on) and the information that the gorse plant with its thorns is closer to the garden pea in the taxonomic classification than it is to the also thorny rose. In these examples the word taxonomy refers to the product, i.e., the finished classification, rather than to the process of classification.

We do not see a major distinction between the terms taxonomy and systematics. Although both are often defined as the 'science of classification', we have chosen to use 'taxonomy' in this book. It has traditionally been associated more closely with the process of classification itself while systematics more often is thought of as including related disciplines like bio-geography and evolutionary biology.

The use of computers in taxonomy is obviously an important component of this book. However, we refer to the recent developments as 'computer-assisted methods' to emphasize that computerized taxonomy is in no way a brand new ballgame. Modern methods of data processing, includ-ing data storage and retrieval as well as analysis, make it possible to utilize a larger quantity of data in a more explicit fashion. Computer assistance can be used to reduce the tedium involved in recording and analyzing large data banks. However, computer-assisted methods should not be expected to re-place the standard methods of doing taxonomy. For this reason the book be-gins with a detailed exposition of what standard taxonomy is and how it is done. We refer to 'standard taxonomy', rather than to traditional or classical taxonomy, precisely to avoid the impression that the time honored methods are somehow outmoded; on the contrary, they serve as the basis for many useful computer-assisted methods.

In addition to providing the taxonomist with practical assis-tance, the computer has had another significant impact on taxonomy. In or-der to develop computer algorithms for doing taxonomic work it was nec-essary to model, in the mathematical sense, the taxonomic process. In many cases this modeling resulted not only in important clarifications with respect to taxonomists' assumptions and thought processes, but also in the devel-opment of new concepts and methods. Inconsistencies were often revealed. For example, taxonomists frequently talk about the need for discontinuities between groups and, at the same time, homogeneity among members of the groups. Use of models indicates that one of these objectives is often achieved at the expense of the other, for a given set of data, and that one must choose an optimal compromise. Greater awareness of the theoretical background and of the explicit bases that underlie the methods allows modern taxonomists both to articulate their activities more clearly and to choose the most appro-priate means for solving practical problems.

The book is divided into four parts. Part I describes the standard methods that have been traditionally used in taxonomy. Part II presents some formal aspects of data structure and the mathematical models that underlie the computer operations used in taxonomy. Part III gives practical details for doing computer-assisted taxonomic analysis—that is, classification, identifi-cation, cladistic interpretation, and so on. Part IV deals with practical means for data storage and information retrieval and gives examples of these meth-ods applied to the large databases that are typical in taxonomy and system-atics.

Throughout the book we have included discussions of the implications of the methods for biological problems in taxonomy like homologies, character selection, natural classification considerations, and so on. A majority of the examples are botanical ones, but we need not apologize for this. Despite the extensive use of computer keymaking for microorganisms, cluster analysis for entomology, and cladistic analysis for vertebrates, it is in angiosperm taxonomy that we find the full range of applications of computers, models, and databases. Character analysis, cluster analysis, ordination, keymaking, on-line identification, parsimony trees, cladograms, compatibility analyses, automated description writing, and databases are all to be found for angiosperms. Of course, the methods and interpretation techniques are applicable to organisms generally. To illustrate this we have included examples dealing with mosquitoes, fish, and even medical diagnosis in human beings.

In places we have used named computer programs as examples. This does not mean that the operating taxonomist should have these particular programs available on the computer. The intention is to describe types of programs so that the reader thoroughly understands the functions a program must perform in relation to one's needs. With this information one can then evaluate the programs available and choose appropriate ones.

Chapter Contents

Part I. Standard Taxonomy

Chapter 1. Taxonomy: An Information System for Biology
The chapter defines taxonomy and its role as an information system serving today's biologists. We consider the direct uses of taxonomy—that is, getting descriptions about taxa by obtaining names and identifications—and the indirect uses of taxonomic data in breeding research, phylogenetic reconstructions, and ecological analyses.

Chapter 2. Taxonomic Products
Before moving into a study of the process of taxonomy, the student should have a clear idea of its goals and products. In this chapter we describe the elements of the standard taxonomic information system. It includes details and samples of monographs, Floras and Faunas, keys, and an introduction to the naming system.

Chapter 3. Standard Taxonomic Analysis
This is an overview of the classification process as it has been done traditionally; that is, how taxonomists have derived sets of comparative characters and used them intuitively to group taxa into hierarchies that are useful and information-preserving.

Part II. Theoretical Bases for Taxonomic Work

Chapter 4. Structure of Taxonomic Data
All studies in comparative biology are based on large quantities of data, which must be marshalled into a suitable form before they can be analyzed. In this chapter we examine the formal aspects of the structure and organization of taxonomic data.

Chapter 5. Models and Algorithms
When computers became available and taxonomists set out to exploit their potential, it was necessary to examine the logical processes which underlay doing taxonomy before useful computer programs could be developed. Models were crucial for clarifying and making explicit the objectives and the methods of taxonomists. In this chapter we present these mathematical models in simple, nontechnical terms.

Part III. Computer-Assisted Taxonomic Analysis

Chapter 6. Character Analysis
The first step in a taxonomic study is to select and define properly the characters that are to be used. Raw descriptors, which are recorded from observations and measurements made on the specimens or taxa under study, are examined by various means for their classificatory potential. The states of each descriptive category that actually occur in the items under study must be ascertained. Since the pattern of distribution of these states is a clue to classificatory usefulness, explicit techniques are described for discerning these patterns and the correlations among them. From these analyses useful characters for taxonomic classifications and keys can be defined.

Chapter 7. Phenetic Classification
The general features of the three major kinds of computerized, phenetic classification methods are described in detail: agglomerative cluster analysis on the basis of similarity measures, optimization cluster analysis, and ordination. The properties of these methods are discussed with respect to particular biological problems and needs.

Chapter 8. Diagrams of Variation Pattern

The computer output from classificatory programs usually requires visual representation for easy interpretation. In this chapter we give practical techniques for both hand-drawn and computer-drawn diagrams, such as scatter diagrams, dendrograms, linkage diagrams, and minimal spanning trees; and we discuss the analysis and interpretation of these displays.

Chapter 9. Identification

Identification aids are an important part of the taxonomic information system. Here we discuss computer programs that will write several kinds of keys and also programs that will make identifications for the user.

Chapter 10. Phylogeny and Cladistics

Here we summarize the four principal cladistic methods used for trying to reconstruct the evolutionary pathway that has led to present-day organisms: narrative methods, methods based on parsimony, cladism, and analysis involving compatibility. While none of these gives reliable results, since the correct evolutionary tree is unknowable, the rapid development of the techniques is of great intellectual interest and promise.

Part IV. Computer-Assisted Database Management

Chapter 11. Database Management

A major contribution made by the computer to taxonomic work is the still largely unexploited area of data storage and retrieval. There are computerized database management systems especially suitable for handling large quantities of data like those in museums and herbaria. Others are better designed for the quick retrieval that is needed for data analysis. Computing facilities, including the use of microcomputers, are discussed.

Chapter 12. Taxonomic Databases

Several specific examples of taxonomic databases that are being used at present are described in some detail. They include curatorial, bibliographic, and descriptive databases, both large and small.

Chapter 13. Goals for the Future

In this final chapter a number of gaps—both theoretical and practical—that remain in the taxonomic information system are discussed along with some dreams for the future.

Approaches to Taxonomy—Our Philosophy

Not all taxonomic work is equivalent. There are several different theoretical bases and inevitable differences in the products that result. With the clarifications resulting from making taxonomy more explicit, fundamental differences have become obvious. Thus, a classification produced by a cladist, whose major objective is deducing phylogenetic trees, will not be expected to be the same as the classification produced by one from the school of numerical phenetics where the emphasis is on computerized cluster analysis based on large numbers of unweighted characters. Both of these will probably differ from classifications produced by traditionally trained taxonomists. In fact the vast majority of taxonomic information comes from the last source; it is quietly produced by practicing taxonomists who do not participate in the arguments that swirl about them regarding the theoretical bases of their working science.

The traditional methods of doing taxonomy are admittedly pragmatic and do not stem from a single, explicit, biologically based theory. Still it seems likely that these methods will be needed and used for some time to come, because the taxonomists' primary responsibility is to provide what is conceptually, if not physically, a comprehensive database—one that maximizes efficiency, accessibility, and generality, and serves many kinds of biologists as well as others.

A major difference between our approach and that of the original numerical taxonomy group, exemplified by Sokal and Sneath (1963, 1973), is related to the matter of character selection. We believe that characters that are useful for making classifications or keys must be carefully refined from the available raw data. Further, careful, considered character analysis and selection is a crucial part of taxonomic analysis that requires a well-trained, experienced taxonomist. Although selecting and defining characters is probably the least well understood of the taxonomic procedures at present, some of the tasks can now be done by verifiable means that are not simply intuitive judgments.

Our view of the process of taxonomic classification, like that of the numerical taxonomists, is essentially phenetic. It seems clear that taxonomic analysis is based on multiple, comparative observations of those traits which can be reasonably assumed to be heritable. The organisms that the taxonomist can observe are necessarily present-day specimens. Since these are the data that the taxonomist must organize, and since we lack direct evidence as to the actual, historical descent of organisms, we feel that the best approach is to form groups of taxa phenetically—that is, on the basis of the relative degree of overall resemblance.

The several means that will be presented for doing phenetic classification all result in a hierarchical arrangement of groups. Hierarchies are efficient means for summarizing information in that features that are invariant at any one level of the hierarchy do not have to be stated repeatedly. They can be assumed once they have been specified at some higher level.

It seems increasingly clear that modern biologists must give up the idea that there will ever be a single, ideal classification for any group. Instead the special purposes of any classification should be clearly stated and the data input and methods of analysis made as explicit as possible. Classification itself should be thought of as a process and not a product. For this reason, classification techniques need to be understood by anyone doing research with comparative biological data, especially where large numbers of variables must be considered simultaneously.

Taxonomy and Computers

The proper role of the computer in taxonomy is a frequent source of controversy. In today's world it is well known that computers are capable of handling "large" data sets "automatically." Taxonomists who use computers need to be aware of two points with respect to this generalization. In the first place, there are practical limits to the size of the data set that can be handled effectively. Whether one uses a large mainframe or a personal microcomputer, there is need for organizing the work with the computer's capacity and other special properties in mind. Biologists using computer centers in universities and other institutions would be well-advised to consult with the staffs of these centers. Microcomputers, on the other hand, are now being designed to be sufficiently "user friendly" so that biologist users are less likely to need the help of experts. Still, they must consider carefully the suitability of their hardware and software for the data they have and the goals they wish to achieve.

A second point that is important to understand is the meaning of 'automatic'. The word suggests machine-like and free from human intervention. Whether this property is an asset or a liability in our context depends upon the taxonomist, and herein lies the controversy. Some taxonomists, notably the numerical school, have believed that automatic analysis by computer ensures against the taxonomist's preconceived biases regarding the proper organization of the taxa and therefore leads to objective classifications that can be produced rapidly and in standardized form. Others resist this view vigorously and scoff at the idea that a trained taxonomist can be replaced by a computer! They argue that no matter how clear the results of a computer analysis or ordination, it is the taxonomist who makes the decisions concerning the classification that is published.

Our philosophy lies between these extremes. As we have indicated above we believe strongly in the potential of computers to assist significantly in the work of taxonomy. Computing machines may well replace some of the activities of taxonomists, but hardly taxonomists themselves. Computerized methods of classification do have the important advantage of being explicit and able to be repeated exactly. The taxonomist will know precisely what has been done and can therefore defend the procedures and decisions specifically. Using computer assistance in this manner enables one to combine expertise in traditional taxonomic methods with the best of computer techniques.

In all fairness we must point out that although the use of computers enables taxonomists to do better work in less time over the long run, it does require a good deal of new work on their part initially. The data must be prepared in standardized form before entry into the computer can even begin. The task of entry is an exacting and time-consuming one, which should include a systematic means of checking for mistakes and correcting errors.

Computerizing data requires a major investment of time—an investment likely to be worthwhile if the result is what we define as a **database**. A database is a collection of data organized to be used by many persons for many purposes; that is, it is planned for multiple or repeated use. As a result of improved computer hardware and software, there are now many coordinated systems of programs in which several analyses can be applied to a single set of data or in which data sets can be merged or joined in various ways. This is particularly true in the case of microcomputers where, for example, word processing programs are now regularly used in conjunction with data retrieval programs to produce reports written in finished form. Thus, multiple use is being facilitated and well-drawn databases are becoming more and more useful.

Part I

Standard Taxonomy

Chapter 1

Taxonomy: An Information System for Biology

The Taxonomic Information System

We consider that the principal objective of taxonomy is to provide an information system, one that provides comparative information about organisms to biologists and to the general public. The first concerted efforts at establishing this kind of systematic information throughout the civilized world were made by Linnaeus (1707–1778). He collated descriptive catalogues, *Species Plantarum* of 1753 and the *Systema Naturae* of 1758, in which plants and animals were classified, described, named, numbered, and provided with means of identification. The need for these orderly presentations came from the rapid discovery of hundreds and then thousands of new organisms brought home as explorers and traders expanded the horizons of the known world.

Linnaeus' classifications belonged to the end of the *a priori* classification tradition and were replaced by the 'natural' classifications of De Jussieus, Adanson, Lamark, and Cuvier in the early nineteenth century. While Linnaeus' classifications were short-lived, his 'information system' as a whole proved exceptionally durable. It laid the foundation of what is still the system today—some 230 years and over a million species later!

The following rather simplistic examples illustrate the three ways in which the taxonomic information system might be used by members of the public today (Bisby 1984a).

Access Through Names

Imagine that a student comes across interesting properties of a named plant in a lecture or a book. He might see a statement like "Many Solanaceae have hallucinogenic properties." Either the name is familiar and conjures up information, however sparse, on the taxon in question—"Solanaceae are tomato-like plants with juicy looking berries"—or the student goes to the library and looks up literature on Solanaceae, possibly following up several references before reaching a description having the required detail in it—perhaps a picture book, or a formal botanical abstract description, or a layman's description in a popular flower book. In any case, the name has given access to the biological information system.

Access Through Published References

In another case, a user may go straight to the taxonomic literature (Floras, Faunas, monographs, or the popular versions thereof) to search out the description and name of a plant with particular properties—an evergreen shrub that will grow near the sea, a ground-cover legume for a sandy garden, etc. The search may be long and tedious; or the user may take shortcuts, like asking for help from an expert or looking at plants similar to ones known to satisfy the requirements. In the latter case the classification is being used in its predictive capacity.

Access Through Identification

Another common way of gaining access to the taxonomic information system is from the desire to identify a specimen. "This curious plant has come up in my garden. What is it and where has it come from?" Possibly one goes to a book (Flora, handbook or monograph) with illustrations and identification keys inside. Depending on one's botanical training and skill in using the technical descriptions that form the clues, the use of a key may or may not result in rapid progress toward a suggestion as to what the plant is. Or one flips through the pages of an illustrated handbook looking for the plant. Or one visits a museum or herbarium and laboriously compares the plant with preserved specimens until a match is found. Whichever method is used, the result is what we call identification—that is, the certainty that the specimen in hand has a particular name. With identification comes the ability to look up other information regarding that plant such as its natural history and its morphological features. The identification has once more led the user into the taxonomic information system.

Upon a little reflection the reader can see that all three of these entry points lead into the descriptive information that is contained in the taxonomic information system. The descriptive data are of many sorts and will serve the needs of many users. A crucial component of their usefulness is their organization. In taxonomy the fundamental element in the organization of descriptive data is the classification, in which specimens or groups are assigned to hierarchically arranged classes in accordance with their overall resemblances. Once a classification is made, the taxonomist can use it as a framework on which to organize the detailed descriptions of the group of organisms. Species and other subgroups are given scientific names, and identification keys are devised to help place individual organisms in their proper groups. The result is the information found in traditional taxonomic publications. This is what we call the taxonomic information system.

Taxonomy in Today's World

Taxonomy is an old science: from the time of Theophrastus' *De Plantis* it has been recognized as basic to biology. A newcomer might think that the tasks of taxonomy must be largely finished. It is true that a usable classification has been available for the major groups of animals and plants since the end of the nineteenth century. The flora of Europe and many parts of North America, for instance, is well known. (It is interesting to note, however, that there is still no book describing all the plants of North America.) Despite this, the task of taxonomy is in reality a never-ending one. Even the simplest catalogue of morphological species descriptions is unavailable for many parts of the world, and if it were to be completed in the next century—which it will not be—there would remain the mammoth, continuing task of revision. Revisions are necessary because people expect the classification and descriptions to incorporate—or at least take note of—an ever enlarging set of additional observations: the presence or absence of certain secondary compounds, newly observed morphological features, breeding data, and so on.

An example of revisionary changes occurs in the placement of the Cactus family (Cataceae). The cacti were formerly classified, on morphological grounds, as an isolated group because of their lack of strong resemblance to other groups. In the 1970s, though, it was discovered that cacti resembled a group of families known as the Centrospermae, in that both possessed the very unusual betalain pigments that are unknown elsewhere. Close examination revealed further resemblances with the Centrospermae families, and that is where the cacti are placed now.

A good example of how improved sampling leads to changes is the case of a rattlepod, *Crotalaria pallida,* found in the humid tropics of the Old and New Worlds. This species was formerly known as three different species, *C. pallida, C. striata,* and *C. mucronata,* each from a different geographic area. It was not until a very extensive survey ended in 1968 that a taxonomist working at the Royal Botanic Gardens Kew realized that all three were actually variants of the same species.

Direct Uses for the Taxonomic Information System

The main purpose of the taxonomic information system, which is often not appreciated, is to provide biological information to people who are not taxonomists. It is because they do provide these services, however, that most taxonomists—working at research institutions, universities, museums, herbaria, and botanic gardens—are supported by public funds. The detailed descriptions of plants and animals and their arrangement in the classification provide the basis for making and testing the hypotheses of the more theoretical aspects of population gentics and evolutionary biology.

Even more important today are the practical needs in areas like conservation, administration, ecology, medicine, law, plant breeding, and biogeography where taxonomic information is essential. Some of these needs are new in that they arise from our twentieth-century problems in dealing with the impact of man on his environment. Others are the old needs biologists have always had for taxonomic products like keys, descriptions, and names. We shall illustrate a number of needs of both kinds in the rest of this section, emphasizing those which are particularly pertinent in today's world.

Example 1

The elementary, but fundamental, need still exists to have unique, unequivocal species names—as well as names for higher taxa—for the organisms being used in any kind of biological research. This is simply a requirement for accurate communication.

Biologists must be wary of using common names, even when they exist (for many organisms have no common names at all). Common names often designate different organisms in different places: witness the English and American robins (members of very different genera in the family Muscicapidae) or the English, Scottish, Texan, and Rocky Mountain uses of the name bluebell for plants in four widely differing families—*Hyacinthoides non-scripta* (Liliaceae), *Campanula rotundifolia* (Campanulaceae), *Clematis pitcheri* (Ranunculaceae), and *Mertensia ciliata* (Boraginaceae).

Much more local sources of name confusion are all too common. In the large lakes of central Ontario, Canada, for example, pickerel is prized as an edible fish while just to the south, in the Great Lakes of the United states, it is considered rather second class for eating. A little investigation reveals that the "pickerel" of the Great Lakes is the "grass pike" of the Canadian lakes, where it is thought to be equally unappetizing.

Common names are also found wanting in that they are necessarily restricted to a single language. For precision and to allow international communication, every scientific publication ought to use scientific names for all organisms mentioned unless no confusion is possible as, for instance, in the case of man in a medical study.

Example 2

Adequate definition of species is more than an academic question in this day of environmental awareness. The governments of both the U.S. and the U.K., as well as of many other countries, have passed laws for the protection of endangered species. Proper enforcement of these laws and certain others requires that administrators, lawyers, and judges comprehend the meaning of the term 'species'. Questions of interpretation arise. For example, to be eligible for endangered species designation, a particular population of fish must be shown to be a taxonomically recognized entity— a species or subspecies—and not simply a minor variation of some other group. Biologists are being called upon to testify in litigations involving questions like this.

Example 3

Plant breeders working on the development of new cultivars of crop plants, with which to feed a hungry world, make use of taxonomic information to choose which plants to use as parents in hybridizations. The breeder needs to know the range of forms available within a given species along with those in other similar species. For example, the delimitation of wild, diploid species of *Solanum* section Tuberosum (relatives of the potato) and the knowledge of cultivated, tetraploid forms are prerequisites for planning breeding programs to produce new disease-resistant and heavy-yielding potato plants. Hence, taxonomic analysis has been an important adjunct to the work carried out by the International Board for Plant Genetic Resources (IBPGR) in building germplasm banks of potato and related wild forms at

Lima, Peru and Birmingham, England. In many germplasm collections a computer database is used for cataloguing the collection and sometimes for containing detailed data that describe and evaluate the holdings.

Example 4

At times biologists make use of the taxonomic information system without actually realizing it. Because of their training in evolution, biologists expect to find similarities among groups that are closely related. An example is the case of plant lectins, proteins that are of interest to pharmaceutical researchers because, among other interesting properties, some of them are specific for human blood groups. One of the early reports of such a lectin indicated that it had been found in the legume *Vicia graminea,* which immediately suggested that other members of the *Vicia* genus should be studied in the search for other lectins and for plants which might produce lectins in larger quantities. Here researchers actually used the taxonomic information system, but probably did it without making a conscious decision to enter the realm of taxonomy.

The above examples are direct uses of the taxonomic information system. We note that several of them exploit not only the summarized forms, like keys and names, but the descriptive information as well. It is the unique responsibility of taxonomists to preserve and codify descriptive data pertaining to specimens and taxa. This information has values far beyond its use in producing classifications. The examples above have already shown that such needs are growing rapidly at the present time. It is fortunate, indeed, that these increased requirements have coincided with the development of computerized data management methods. A major development in taxonomy is currently underway in the form of computer-assisted data storage and retrieval to supplement the traditional methods of disseminating descriptive data via monographs and museum catalogues.

Other Uses of the Taxonomic Information System

In addition to the direct uses of the taxonomic information system there are a number of higher level and interpretive uses for the information. Some of these, such as phylogenetic reconstruction and vicariance biogeography, are traditionally associated with taxonomy: together with taxonomy they form the wider discipline of systematics. Other interpretive uses of taxonomic information are seen in the adaptive and evolutionary strategies devised by ecologists and evolutionists. All of these depend heavily on the taxonomic information system, both for framing problems and for data with which to solve them.

Phylogenetic reconstruction bears a controversial relation to taxonomy. We believe that the connection between taxonomic and evolutionary groups is by no means automatic and that it is important to understand that evolutionary interpretations are not the primary purpose of much work in taxonomy. Some would agree with us that the various aspects of reconstruction, such as deduction of possible cladistic trees showing evolution's route and deduction of evolutionary timing and rates of change, are secondary studies that utilize the classification as a starting point. There are others, with whom we disagree, who believe that a phylogenetic reconstruction, no matter how speculative, should be used to shape the final classification and information system. This is supposed to give the classification evolutionary properties and naturalness, but in our opinion it introduces additional problems, and in practice is an extra source of instability.

Researchers in agriculture are beginning to apply the methods of taxonomic analysis to their databases for cultivars—i.e., plant variants maintained by and for cultivation. In addition to the traditional taxonomic information that workers like the plant breeders in example 3 above use, agricultural experience generates a large amount of information on cultivars and their special qualities and requirements. Workers in agricultural research are often interested in relationships and "natural" groupings among their cultivars and desire to classify them. The newly developed multivariate classification programs are now being used for agronomic data.

An example of a classification for a special purpose, which uses agronomic data, is the study of Edye, Williams and Pritchard (1970) in Australia on the pasture legume *Glycine wightii*. Since Australia has a depauperate flora, there is great interest there in introducing useful plants. This *Glycine* species, introduced from Asia and Africa, was analyzed by classification "to evaluate the relative agronomic performance of introductions and their taxonomic affinities . . . [to] lead to more effective collection, introduction, and evaluation of further introductions of the species." The analysis was made on the basis of 31 characters, some morphological like those in many standard taxonomic information systems, and others that were agronomic features of the plant populations examined. The results were six groupings of cultivars, which could be examined in relation to their particular agricultural uses.

This study illustrates two general features of special purpose classification: first, that the criteria for the formation of the groupings are sometimes arbitrary in the sense that agricultural classification criteria need be relevant only to a particular man-made problem. Second, special-purpose

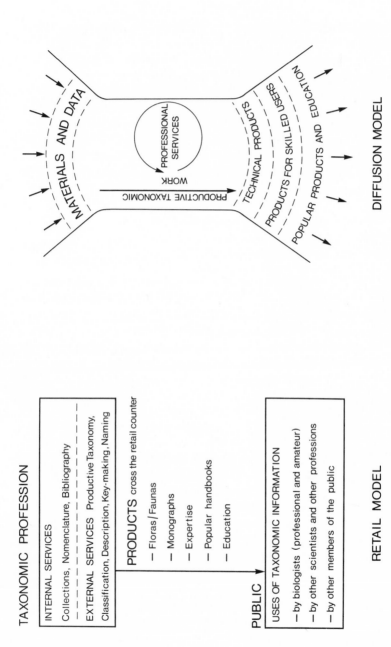

Figure 1.1 The retail (a) and diffusion (b) models of information services provided by taxonomists. *(From Bisby 1984a, fig. 1 and 2 respectively)*

classification in practical situations often requires that the resulting groups be diagnosable in terms of the attributes that characterize them. This is not a general requirement in taxonomic classifying, except for characters for keys.

Plant ecologists also make wide use of some methods that are derived from taxonomic analysis. Their data are often that of species associations combined with environmental or morphological attributes. They then use methods of the classification type. For example, an ordination is used to produce plant groupings based on habitat resemblance, which can then be related to environmental gradients. (See Clifford and Stephenson [1975] for a full discussion of ecological classification.)

Models for Taxonomy as a Whole

To sum up, we think of taxonomy as providing an information system for biology. As a subject it raises very many fascinating theoretical questions, as well as providing the great interest of working with natural variation, but ultimately its value lies in the service element: does it meet the needs of the user, people at large who are not taxonomists.

We illustrate this view with two simplistic models from Bisby 1984a. The Retail Model (fig. 1.1a) emphasizes that the real products of taxonomy are the elements which ''cross the retail counter'' to customers. Professional collections, work on nomenclature, and collation of bibliographies are thought of as internal services that support the external services. The Diffusion Model (fig. 1.1b) is more realistic in that the edges of taxonomy are blurred: information of taxonomic value comes in from the rest of biology but the dividing line is imprecise. And similarly the products of taxonomy—classifications, nomenclature, descriptions, and identification aids—diffuse out from the technical taxonomic publications and move eventually into popular knowledge.

Suggested Readings

Elementary Introductions to Taxonomy

Heywood, V.H. 1976. *Plant Taxonomy*, 2nd Edition. London: Arnold.
Jeffrey, C. 1968. *Introduction to Plant Taxonomy*. London: Churchill.
Savory, T. 1970. *Animal Taxonomy*. London: Arnold.

Chapter 2

Taxonomic Products

As we have established in the last chapter, a major role of the taxonomic profession is to provide a taxonomic information system. The system centers around the hierarchical classification of organisms. There are technical forms used for this information. **Floras** and **Faunas** are descriptive catalogues or manuals of plants and animals respectively, usually for a restricted geographic area. Taxonomic **monographs** (also referred to as revisions) contain the classification and identification keys for a taxonomic group such as a genus or family, along with the comparative data, in brief form, which were used in the preparation of the classification. Taxonomic information is also available in other forms ranging from museum collections to popular handbooks. Descriptions of these taxonomic end products, so that the purposes and useful results of taxonomic work will be clear to the reader, are provided here.

There are four important elements of the taxonomic information system: the **classification,** a hierarchical arrangement of taxa; the **descriptions** for each level; the **nomenclature,** an internationally accepted set of Latinized names; and the **identification aids,** such as keys, pictures, collections, and other standards for comparison. Of course, these four major elements are interrelated, but for simplicity we will treat them one at a time. The classification is the central element although it is abstract and not as apparent to the public as the nomenclature or the identification aids. However, the latter are in fact dependent on the existence and substance of the classification.

The Classification

Hierarchical grouping of organisms is the system of biological classification in almost universal use. The groups at any level in the hierarchy are called **taxa** (singular **taxon**). At each level the set of taxa is both exhaustive and exclusive; that is, every organism included in the classification must belong to one of the taxa and can not belong to more than one. For example, the families Leguminosae and Rosaceae are both members of the order Rosales—and of no other. Similarly, each taxon contains taxa of lower rank nested within it: the bumble bee genus, *Bombus,* comprises, among others, the species *Bombus lapidarius* and *Bombus terrestris.*

A fixed list of rank or level names, known as categories, is used throughout botany, zoology, microbiology, paleobotany, agriculture, and horticulture. It is given below. The names read down from the higher, more inclusive, ranks to the lower, more specific, ones. At any one rank the member taxa are all equivalent; for example, at the level of family every taxon is a family. It should be clear that the taxa at the higher levels are more diversified and contain larger numbers of organisms.

Categories

Kingdom
Division or Phylum (plural: Phyla)
Class
Order
Family
Tribe
Genus (plural: Genera)
Section
Species
Subspecies
    (Variety, in plants)
    (Cultivar, in plants)

The categories of the hierarchy employed in animal, plant, and bacterial classifications are generally equivalent, although the category superfamily or family-group is used in some animal phyla. There are sometimes differences in the names of categories used, as is the case for plants and bacteria where Division is used in place of Phylum. The names of the categories used for subdivisions of the species vary among plant and animal groups.

The extent to which subcategories (such as subfamily) are used also varies. Generally zoologists use more subdivisions between the ranks of phylum and genus and botanists use a greater number of subdivisions be-

neath the rank of genus. Whether these differences reflect genuine differences in plant and animal variation patterns or whether they reflect the isolation of plant and animal taxonomists in the late nineteenth and twentieth
centuries is uncertain. As we shall see later, there are few objective criteria
for determining the rank of higher taxa, such as groups of species, and even
the attention given to species definition has failed to provide uniformity or
total objectivity.

The criteria that form the bases of the classification are matters
of considerable importance; a classification is not mere haphazard pigeon-
holing. In the past people used **artificial classifications** in which organisms
were arbitrarily positioned by a few characters selected on *a priori* grounds.
For example, uses and healing properties of plants were used in classical
times and in his so called Sexual System Linnaeus used the number of stamens and pistils in the flower.

Despite Linnaeus's unrivaled achievement in stabilizing nomenclature and setting up a fixed plant classification in the late eighteenth century, the Sexual System came under increasing criticism, particularly by Pa-
risian biologists, in the early nineteenth century. In their different ways Bernard
de Jussieu and Adanson in Paris, and to some extent de Tournefort and Ray
before them, argued that the classification should reflect the natural pattern
of resemblances based on observation of a wide range of homologous characters from several parts of the plant. Each of the characters observed contributes an element of pattern, of resemblances and dissimilarities, to the
collective or "natural" pattern. At that time the characters were all from
observations of gross morphology but today we extend the concept to the
natural pattern among whatever characters we use. **Natural classifications,**
which reflected these natural patterns, gradually superseded the Linnaean
classification during the nineteenth century.

A further, more recent, complication has been the introduction
of another concept of natural pattern—that suggested by evolution and common descent rather than by overall resemblance. Differences between these
two uses of the term 'natural pattern' are best resolved by referring to the
two modern terms 'phenetic' and 'phyletic'. **Phenetic natural patterns,** like
the earlier concept, are based on resemblances; **phyletic natural patterns**
are based on the more recent concept, stemming from descent.

There are some parts of classifications produced even today that
are artificial, and yet others that incorporate hypotheses about the route of
descent. Most botanical classifications, however, reflect the phenetic, natural pattern. They reflect estimates of overall resemblance or similarity—such
as that daisies are more similar to sunflowers than to roses—and give groupings that show discontinuities between and continuity within groups. The
taxonomist producing these classifications must observe the pattern of re-

semblances and reflect this in the classification. As we shall see, such awareness of the natural groupings can be thought of as pattern analysis, although probably few practicing taxonomists would express it this way.

Conversely, someone unfamiliar with a group can use the structure of the classification to deduce features of the pattern. Very isolated organisms are placed in monotypic groups—that is, the next higher taxon contains only one taxon within it. An extreme example is the coelocanth, a primitive fish, which is placed in a monotypic genus, in a montoypic family, and so on up to a monotypic subclass adjacent to the subclass Teleostii—which has, by contrast, hundreds of species. The maidenhair tree, *Ginkgo biloba,* is similarly isolated as the only member of the order Ginkgoales in the Gymnospermae.

Because similar species are classified into the same sections, and similar sections into the same genus and so on, we can tell from the distribution of species and sections in the genera of a family what the natural pattern is like. For instance, the Umbelliferae and Myrtaceae are both families of approximately the same number of species (2900 and 3000). However, since the Umbelliferae has its species fairly evenly divided into 200 or so small genera, we can tell that its species form a diverse pattern of very many small groupings. By contrast, the Myrtaceae consist of 80 genera, 8 of them with from 100 to 1000 species. Here the pattern is different: a few large associations of species and a much smaller list of small groupings.

The Descriptions

The hierarchical classification provides a structure for recording information at any level. Each taxon has both a name and some descriptive generalizations. These generalizations are entered at the rank where they apply to all the organisms belonging to one of the taxa, but ideally to none of the others at the same rank. So, for instance, the occurrence of seeds enclosed in ovaries and xylem vessels in the conductive tissues is recorded as a generalization for the class Angiospermae. These features characterize all of its members, but not the members of Gymnospermae. It would be of little use to record these characteristics at the higher level, the Division Spermatophyta, because they do not apply to all members of the division. Nor would this information be usefully recorded at the lower family level, where it would have to be repeated for each of the several hundred families that have seeds in ovaries and xylem vessels. As a consequence of such organization, the descriptive information is efficiently arranged. Note that information is much

more general for taxa at higher ranks than at lower ranks. The Euphorbiaceae, or spurge family, is divided into about 300 genera. Clearly the description of the family will allow for more variation among its members than
the description of *Manihot,* or cassava, which is one of the genera within it.

The descriptions of taxa in Floras, Faunas, and monographs, at
whatever hierarchical level, are in essence abstracts. Usually the descriptions are given in telegraphic style to save space and to make finding a particular piece of information easier. None of the descriptions can be said to
be complete. These abstracts usually give only certain types of data, frequently that for the external morphology, and contain only the kinds of descriptive data that the author can honestly claim to have observed directly.
In nearly all cases there is a definite form in which the abstract occurs. While
it may make dry reading, it is not intended to do more than present the data
in a manner that allows for the comparison of one taxon with another.

Fig. 2.1 shows a page from a Flora (Small 1933). Note the three
different levels of the hierarchy (family, genus, and species) and how each
bears a name and a description. A key to the genera is also found in this
example. Much of the description is in note form, each sentence consisting
of a noun followed by a list of adjectives or adjectival phrases applying to
it. So, for example, in the first line for Aceraceae, "Leaves opposite: blades
simple or compound" is note form for "Leaves are in pairs whose members
are opposite each other on the branches, and the leaf-blades are either simple
or compound."

The technical form of many Floras, Faunas, and monographs
unfortunately makes them unsuitable for use by many amateur natural his-
torians and by scientists in other disciplines. In many cases these people's
needs are met by the large body of popular literature, such as handbooks or
field guides, that describes organisms in more ordinary language and makes
much more use of illustrations. An example is given in fig. 2.2, where the
entry for *Acer* species in Nelson (1969) can be compared with the description of *Acer* shown in fig. 2.1. The example also illustrates the updating that
goes on constantly in taxonomy. Note that by 1969 the *Acer* classification
had been revised and, for example, *Acer negundo,* the boxelder of North
America, had the epithet *negundo* taken from the earlier treatment in which
this species was in a separate genus called *Negundo.*

In some cases information in popular publications derives directly from a technical publication. Thus, Polunin (1969) acknowledges that
he has drawn heavily on material from the condensed, technical *Flora Europaea* (Tutin et al. 1964–1980). It should be stressed that whereas most
popular books give simple descriptions, good illustrations, and easier identification of common species, they are often unreliable for precise identification and description. This is due not to inaccuracy, but—in nearly all cases—

FAMILY 11. **ACERACEAE** — MAPLE FAMILY

Shrubs or trees. Leaves opposite: blades simple or compound. Flowers perfect or polygamous, in cymes, racemes, or panicles, or often in congested clusters. Calyx of 4 or 5, or rarely more, deciduous sepals. Corolla of 4 or 5, or rarely more, petals, or wanting. Androecium of as many stamens as there are sepals or twice as many. Gynoecium of 2 more or less united carpels. Fruit 2 nutlets with wings (samaras).—Six genera and more than 100 species, in the north temperate zone.

Leaf-blades simple: flowers polygamous, monoecious, andromonoecious, or androdioecious; disk present: anthers ellipsoid or oval, not tipped.
  Flowers in terminal racemes or panicles: stigmas shorter
    than the style.   1. ACER.
  Flowers in lateral or terminal clusters: stigmas as long
    as the style or longer.
    Flowers filiform-pedicelled, in drooping clusters appearing with the leaves: sepals united into a lobed cup-like calyx, the staminate and pistillate similar.   2. SACCHARODENDRON.
    Flowers sessile or short-pedicelled, in dense lateral involucrate clusters, appearing before the leaves.
      Sepals united, the staminate and pistillate calyx
        very distinct: petals wanting.   3. ARGENTACER.
      Sepals distinct, those of the staminate and pistillate flowers similar: petals present.   4. RUFACER.
Leaf-blades pinnately compound: flowers dioecious: disk wanting: anthers linear, minutely tipped.   5. NEGUNDO.

1. ACER L. Shrubs or trees. Leaf-blades broad, coarsely toothed or 3–5-lobed. Flowers borne in terminal racemes or panicles, appearing after the leaves, polygamous. Calyx of usually 5 distinct or slightly united sepals. Petals 5, narrower or broader than the sepals. Stamens exserted or included: stigmas shorter than the style.—About 30 species, North American and Eurasian.—Spr.—MAPLES. ACERS.

Flowers in erect panicles: petals linear or linear-spatulate, about twice as long as the sepals.   1. *A. spicatum.*
Flowers in drooping racemes: petals obovate, about as long as the sepals or slightly longer.   2. *A. pennsylvanicum.*

1. A. spicatum Lam. Shrub, or small tree 10 m. tall, the bark thin, relatively smooth: leaf-blades mostly longer than broad, mainly 3-lobed, sometimes with 2 additional lobes near the base, serrate, glabrate above, paler and more or less tomentulose beneath, cordate or subcordate: panicles many-flowered: pedicels spreading, 6–10 mm. long, or longer at maturity: petals yellow: stamens exserted: fruit green, about 2.5–3.5 cm. broad, the wings of the samaras spreading at about 90 degrees.—(MOUNTAIN-MAPLE. LOW-MAPLE.)—Damp rocky woods, Blue Ridge and more northern provinces, Ga. to Man. and Newf.—Spr.–early sum.—The heart-wood is soft, and light.

2. A. pennsylvanicum L. Shrub or tree, rarely over 11 m. tall, the bark relatively smooth, longitudinally striped: leaf-blades

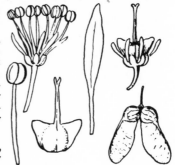

Figure 2.1 Portion of a technical description of the Maple family. (From Manual of the Southeastern Flora *by John Kunkel Small. Copyright 1933 The University of North Carolina Press. Used with permission of the publisher).*

MAPLE FAMILY, *Aceraceae*

This is a family of trees and shrubs native in the cool regions of the northern hemisphere. All North American species belong to the genus *Acer*. It is characterized by opposite, and usually palmately lobed, leaves and winged and paired fruits. Three species occur in our area.

ROCKY MOUNTAIN MAPLE, *A. glabrum* (189), is a large shrub or small tree with gray bark. Its young twigs are smooth and dark red and its winter buds are bright red. The leaves are palmately 3- to 5-lobed and sometimes 3-foliate, always sharply toothed. Its small chartreuse-colored flowers are fragrant and in the center of the female ones you can see the young ovary already in the shape of the double "key" which is the maple fruit. As these ripen they are often tinged with red. There is an insect which frequently attacks this maple, its egg clusters appearing as crimson blotches on the leaves. It always grows with several trunks in a clump. These may be from ½-inch to 5 inches in diameter and from 6 to 30 feet tall. It is found from Wyoming and Idaho south to Nebraska, New Mexico and Utah and is common in the canyons and on hillsides of the foothill and montane zones throughout Colorado.

189. ROCKY MOUNTAIN MAPLE, 2/5 ×

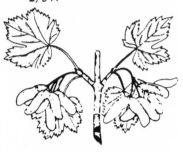

WASATCH or BIGTOOTH MAPLE, *A. grandidentatum* (190), is a small tree which grows in groves or thickets on moist canyon sides and along streams. Its leaves are 3-lobed with rounded sinuses and bluntly pointed tips. They turn rose or red in autumn and some seasons make a very spectacular color display. This species is found from western Wyoming through the mountainous parts of Utah to Arizona, New Mexico and west Texas.

BOXELDER, *A. negundo*, is a tree found along streams of the high plains and foothill canyons. It differs from our other maples in having a pinnately compound leaf and flowers and fruits in pendant clusters.

190. WASATCH MAPLE, 2/5 ×

Figure 2.2 Part of a popular description of the Maple family. *(From Nelson 1969:190 & 191)*

to the need to omit species because of space limitations. In addition, popular books cannot provide illustrations or photographs of sufficient quality to distinguish among similar species.

The Nomenclature

Each taxon bears one scientific name in addition to any popular or common names. Regardless of any intrinsic meaning it may have, the name has an absolutely essential function in communication and retrieval of biological information. It is used as a noun in communication; it provides one of the entry points to the taxonomic information system; and it acts as a tag for cross-referencing within the system. Clearly the name must be distinctive and be unambiguously attached to one taxon.

The scientific name for each taxon is in Latinized form, often having a distinctive ending or arrangement that is appropriate to a particular category. Thus, Apiidae can be recognized as an animal family name by the ending '-idae', and *Crotalaria nitens* can be recognized as a species by the binomial form (discussed below).

By international agreement taxonomists who name taxa follow the codes of nomenclature: the International Code of Zoological Nomenclature for animals, the International Code of Botanical Nomenclature for plants (including fungi), and the International Code of Nomenclature of Bacteria for bacteria and actinomycetes. There are well established procedures for modifications of the rules. In botany, for example, committees of the International Botanical Congress consider proposals on rules and submit their recommendations to the Congress, which meets every five years and which then publishes any changes in the rules. The International Code of Nomenclature of Cultivated Plants, in connection with various national laws, governs their naming. An outline of how names are published is given in chapter 3, and for an excellent summary of the codes, see Jeffrey (1977).

The names of genera are single words, Latinized nouns with capital initial letters, usually underlined in handwriting and italicized in print. Examples are shown below.

Names of Genera
Drosophila (the fruit flies)
Quercus (the oaks)
Lupinus (the lupines)
Bovus (the cows)
Fuchsia (a genus named after Fuchs)

A few generic names *(Quercus, Bovus, Triticum, Equus)* are the common names once used by the Romans, but most are invented or contrived from descriptive elements, names of biologists, or are literary allusions.

The names of species are **binomials,** that is, they consist of two parts. The first part is the name of the genus to which the species belongs and the second is the Latinized specific name or **epithet.** The combination of the generic name with the epithet is unique, although the epithet itself may be used in other genera as shown in the examples below.

Names of Species
Crotalaria juncea (sunn hemp plant)
Inga edulis (edible inga)
Ostrea edulis (oyster)
Spartium junceum (Spanish broom)
Passiflora edulis (passion fruit)

The epithet is commonly an adjective, and must agree with the generic name in gender. Thus, in the list above, the adjective *juncea* (meaning rush-like) agrees with *Crotalaria,* which is feminine, while *junceum* is used with *Spartium,* which is neuter. The epithet may be an exact repeat of the generic name only in bacteria and animals, where, for example, *Bison bison* is the scientific name for the American buffalo.

The species name must always contain the two parts, but to save space when repeated mention of species from the same genus occurs, often only the initial letter of the generic name is used. The epithet is usually given a small initial letter although it was once the habit to use a capital if it contained a proper name. (Using either generic or specific names to commemorate the names of biologists is common.) Generic names and species binomials are underlined in hand- and type-written texts and italicized in print. Example:

Ilex cuthbertii = *I. cuthbertii* = *I. Cuthbertii*

Subdivisions of a species (known as infraspecific taxa) have **trinomials** as their correct scientific names. The first two of these names are always the binomial of the species to which they belong. In the case of plants or animals from the wild the third name is a Latinized name peculiar to the taxon, as in the following examples.

Cervus elaphus canadensis (the N. American elk, subspecies of the species called red deer in Britain.)
Dactylorhiza fuchsii subsp. *okellyi* (a white ecotype of an orchid, peculiar to the southwest of Ireland)
Retama raetam var. *sarcocarpa* (a wild, coastal variety of a Mediterranean species of broom)

The first of these three examples is the correct form for an animal subspecies—i.e., three names with no designator. The second (a fine example of Latinized surnames) and the third examples are of a **subspecies** and a **variety** of plants. It is essential here to include the designator 'subsp.' or 'var.' to show which is which. The variety is a category of lower rank than the subspecies. Officially the **form** exists as a third category of even lower rank; it is designated by 'f.' or 'forma', but is rarely used. Sometimes a plant subspecies is further subdivided into varieties but it is often unnecessary to write out the name in full, as in *Vicia faba* subsp. *faba* var. *minor* (the tick bean) which can be referred to as *Vicia faba* var. *minor*.

Manmade cultivars, which tend to be known familiarly, and confusingly, as varieties by many plant breeders, horticulturists, and agronomists, bear the name of the species to which they belong. The third name must, unlike all other scientific names, be in a modern language rather than in Latinized form. It can be written in two ways, with the designator "cv." or with the name in quotation marks. Thus,

Pisum sativum cv. Sugar Baby or *Pisum sativum* 'Sugar Baby'.

Cultivar names like those above may be translated or even substituted for when the cultivar is offered for trade in another country. However, an effort has been made in drawing up the EEC countries' *Common Catalogue* to locate synonyms and require that seeds and other materials be sold under one of the names only. Further regulation of cultivars is controlled by plant breeders' rights and cultivar registration laws and practices in the various countries. Plant breeders' rights, conferred on the person who registers a new cultivar in the correct way, are broadly similar to copyrights or patents.

Genera and categories above the generic level have names that are single, Latinized words. Names of many categories have prescribed endings. The 'idae' ending for animal families was mentioned above. Plant families usually end in 'aceae'. However, there are eight families of flowering plants—unfortunately families which are very frequently encountered—for which older names ending in 'ae' are still accepted alternatives. Some of these common older forms, and their newer alternatives, are: Palmae or Arecaceae (palms), Cruciferae or Brassicaceae (cabbage family), Gramineae or Poaceae (grasses), Leguminoseae or Fabaceae (pea family), Compositae or Asteraceae (daisy family).

This example illustrates the fact that we have a legacy of names given before the formal nomenclatural rules were in place. Formal nomenclature as we now know it began with the works of Linnaeus in the eighteenth century, and gradually became standardized in the early twentieth century. Consequently there are sometimes difficulties with older names for which the details of publication, description, and citation of types may be

missing or obscure. However, because of various rules, including the rules
of priority, they can not be ignored.

In technical taxonomic publications, scientific names of organisms are often followed by a sometimes abbreviated **authority,** the name of
the author who originally published the name, as in *Spartium junceum* L.,
published by Linnaeus, or *Lathyrus arizonicus* Britton, meaning published
by Britton. This precision is particularly useful, where, by accident, the same
name has been used twice in breach of the nomenclatural rules. For example, on two occasions different plants have been named *Lathyrus sessilifolius.* By referring to the authority we can distinguish between them. Thus,
the plants named as *Laythyrus sessilifolius* Tenore (with Tenore as the authority) are now included in *Lathyrus digitatus,* whereas plants named *Lathyrus sessilifolius* Hook. & Arn. (with Hook. & Arn., as the authority) are
now included in another species, *Lathyrus hookeri.*

If the organisms in a given taxon have, at some time, been known
by one or more other names, the other names become **synonyms** of the accepted names. Changes in names occur either for genuine biological reasons
(these are **taxonomic changes**) or to meet the naming rules (**nomenclatural
changes**). A taxonomic reason might be the merging of species into one in
a revision on the basis of new data, whereas a nomenclatural reason might
be the discovery that a name is illegitimate or invalidly published for any
one of a variety of technical reasons. According to the rule of priority, the
earliest legitimate, validly published name (at least back to the various starting dates) is the correct name. So in the examples of the paragraph above
versions of the name *Lathyrus sessilifolius* have become synonyms, often
listed thus:

> *Lathyrus digitatus* Fiori
>   (syn. *L. sessilifolius* Tenore)
> *Lathyrus hookeri* G. Don
>   (syn. *L. sessilifolius* Hook. & Arn.)

The Identification Aids

Identification is the act of determining where a specimen belongs in a given
classification. A student who takes a specimen of the redbud or Judas tree,
Cercis, which is new to him, and finds that its position in the classification
is within the family Leguminosae is said to have identified the specimen to
the family level. This activity is important not only because it provides a

name for the unknown specimen but also because this name provides a link with the information stored in Floras about this plant and its relatives. Of course, identifications can be made at any taxonomic rank; in biological work identification is commonly carried to the species level.

Identification and classification are frequently confused. What taxonomists call identification, as defined above, is sometimes called classification by statisticians. Also some biologists, whose only conscious contact with taxonomy is in identifying specimens, do not understand that taxonomic classifications have objectives other than identification.

Usually identification is done by utilizing a printed structure called a **key.** Keys cannot be made for a particular group of organisms until there is a classification. They may be made for identification at any taxonomic level; there may be a key for the species of a genus or for the genera of a family. Using a key requires the consideration of a series of **dichotomies.** These are in the form of pairs of contrasting, descriptive statements called **leads,** alternative statements between which one must choose at each step.

As an example, consider a key included on the page from a Flora (Small 1933) shown in fig. 2.1. (Keys on this page are shown in the smallest print and there are two of them.) Here, the first dichotomy is "Leaf-blades simple" contrasted with "Leaf-blades pinnately compound." The user expects to find that only one of the leads is correct for the specimen being examined. Thus, if leaf-blades are observed to be simple, this eliminates the possibility that the specimen belongs to the *Negundo* group, which have pinnately compound leaves. The layout of the key is such that the user proceeds from this first pair of leads to another dichotomy, where the observation and decision process is repeated and again some of the possibilities are eliminated. The identification is completed by working through this elimination process until finally the key leads to a single name. That is the answer.

The simplest key layout to use is the **bracketed key,** where numbered leads are printed one above the other and the directions to later dichotomies or answers are on the right. Thus the Aceraceae key from fig. 2.1 could be printed as follows:

1a. Leaf-blades simple: flowers polygamous, monoecious,
     andromonoecious, or androdioecious; etc. _ _ _ _ _ _ _ _ _ _ _ _ _ _ _ 2.
1b. Leaf-blades pinnately compound: flowers dioecious; etc. _ _ 5. *Negundo*
2a. Flowers in terminal racemes or panicles; etc. _ _ _ _ _ _ _ _ _ _ _1. *Acer*
2b. Flowers in lateral or terminal clusters; etc. _ _ _ _ _ _ _ _ _ _ _ _ _ 3.
3a. Flowers filiform-pedicelled, etc. _ _ _ _ _ _ _ _ _ _ _ 2. *Saccharodendron*
3b. Flowers sessile or short-pedicelled, etc. _ _ _ _ _ _ _ _ _ _ _ _ _ _ _ 4.

Here "leaf-blades simple" directs the user to dichotomy number 2. If the flowers are in racemes or panicles, lead 2a directs the user to an answer, the genus *Acer*. If the flowers are in clusters, lead 2b directs him to dichotomy number 3 and so on.

The **indented key** such as the one in fig. 2.1 is an alternative form of key layout. In this particular example the leads are not numbered and the user has to spot the pairs from the degree of indentation from the left and from the matching initial words. Also there is no numbering to direct the user to the next dichotomy. Instead a pair of leads may be separated, as with the first dichotomy in the example. If the first lead is chosen, the user takes the dichotomy that starts immediately below and is indented one place to the right. If the second lead is chosen, subsequent dichotomies would be found beneath it. Whenever an answer is found, the name of the group indicated is printed on the right.

Identification is only half completed once a name has been found in a key. The other half of the job that must be done is to confirm the identification by checking to see that the specimen fits the descriptions and illustrations. Previously identified specimens in herbarium or museum collections can also aid in verification of the identification. The characters listed for the putative taxon, particularly those characters that have not been used in the key, are useful as checks.

The key we have been using as an example is of a generic key. Following the description of the genus *Acer* in figure 2.1 the author places a second key. It allows identification of the unknown specimen to the level of species, either *A. spicatum* or *A. pennsylvanicum*. This practice of providing separate keys at each level of the hierarchy is common. It means that keys are short and that the user alternates between keying and reading descriptions. Further, artificial keys—those that do not parallel the natural classification—are often more efficient, so that it is not always advantageous to key out by going through the taxonomic hierarchy layer by layer. Also note that important descriptive data are often contained in the leads of the key; they can add substantially to the user's knowledge of a taxon.

Suggested Readings

General

Radford, A.E., W.C. Dickison, J.R. Massey, and C.R. Bell 1974. *Vascular Plant Systematics*. New York: Harper and Row.

Simpson, G.G. 1961. *Principles of Animal Taxonomy*. New York: Columbia University Press.

Nomenclature and Identification

Jeffrey, C. 1977. *Biological Nomenclature*, 2nd Edition. London: Arnold.
Pankhurst, R.J. 1978. *Biological Identification*. London: Arnold.

Discussion

Bisby, F.A. 1984. Information services in taxonomy. In R. Allkin and F.A. Bisby eds. *Databases in Systematics*, pp. 17–33. London and New York: Academic Press.
Heywood, V.H. 1984. Designing Floras for the future. In V.H. Heywood and D.M. Moore eds. *Current Concepts in Plant Taxonomy*, pp. 397–410. London and New York: Academic Press.
Prance, G.T. 1984. Completing the inventory. In Heywood and Moore eds., pp. 365–396.

Chapter 3

Standard
Taxonomic Analysis

N ow that we have described the taxonomic information system, the product of the taxonomist's work, we turn to a consideration of the processes by which it is produced. The monographic revision is central to the process. It is a survey of a taxonomic group that is aimed at making a fresh classification because an earlier study of the same taxa has proved to be incomplete or inaccurate. Completion of the new classification is followed by revision of the information system's descriptions, nomenclature, and keys. Ideally the monographic revision, be it of a species, genus, tribe, or family, should include all known or suspected members on a worldwide basis. Although this is sometimes possible for small groups that inhabit restricted areas, practical limitations set by time, money, materials, and even politics often mean that monographs are based on something less than the entire membership for large or widespread groups.

The other principal kind of taxonomic work, Flora and Fauna writing, will be largely omitted here. It is evident that in an ideal world, where monographs would be available for all of the world's taxa (actually a goal of near impossible magnitude but see Stace [1980: 246] for calculations), writing Floras and Faunas for the continents and more detailed ones for smaller regions would be a matter of collation, editing, and key-writing rather than critical classificatory work. In practice things are very different. Fauna and Flora writers often do critical classificatory work on species for which there are no monographs as well as for species that are covered.

The Flora Neotropica and Flora Europaea projects illustrate both extremes. In the latter case, the majority of European species were already well-known when Flora Europaea was started. So the Flora writing was a job (albeit a mammoth one!) of getting a classification agreed to by the specialists and of preparing keys, descriptions, and national distribution lists—all for reasonably well known species. The task took a team of editors, a secretariat, and a panel of national experts about 25 years to complete. In contrast, the Flora Neotropica project is charting much unknown territory. The majority of groups in the enormous flora of the American tropics are without monographs and indeed are very poorly represented by materials in herbaria. Hence, the Flora Neotropica is essentially a series of monographs for groups whose center of diversity is in the tropics. Much collecting and critical classificatory work is needed and progress has been slow. Estimates of the time needed to complete the Flora Neotropica range far into the next century.

Monographic revisions can be made at a variety of taxonomic levels and at various stages in the taxonomic history of a group. The first stage, often known as alpha taxonomy, is that of basic recognition and description of species. It involves surveying morphological materials and describing the species, usually on the basis of scant information on the variation pattern and without any breeding data. As knowledge of the group accumulates it becomes possible to introduce more biological, chemical, and breeding data into the analysis. As this is done the study moves further along the scale toward omega taxonomy (Turrill 1935). It also becomes easier, once the alpha taxonomy is complete, to move up the hierarchy and survey the species and genera of a tribe and the tribes of a family. Any particular monographic project comes at a known stage in the history of a taxonomic group, in that there are by then declared upper and lower limits. Take for example Kupicha's survey of the tribe Vicieae (Kupicha 1976). In it she is concerned with the classification of species into groupings within the tribe. Her attention is focused on groupings between these limits—i.e., the sections, subgenera, and genera into which her material falls.

What follows here is a generalized description of the steps in a monographic survey. It suffers, of course, from this generalization. Details will vary with position on the scale from alpha to omega taxonomy, they will vary with taxonomic level, and most of all they will vary with the biological and taxonomic details peculiar to each group. Each taxon is unique, and this provides some of the fascination of taxonomy.

The steps in taxonomy parallel those followed in other sciences and we shall describe them in the traditional categories that are used to report research: 'Introduction', 'Materials and Methods', 'Results', 'Conclusions'. Upon a little reflection the reader will see that the first step is basically a planning task, the second is the active research phase, the third is the time for data analysis, and the last for formulation and publication of the new body of knowledge.

Applying these four general steps in a more specifically taxonomic context we arrive at the following flow of procedures which we will develop in detail in the rest of this chapter:

Introduction. The first step in any project is to define one's problem in a form that delimits its scope and makes it potentially solvable. One must come up with a working definition of the problem in the beginning, although the expectation is that the exact definition of the limits of a study will be reformulated during the literature survey, which is a major activity during the introductory phase of a monographic study. While searching bibliographic sources to find the names and descriptions of the taxa, taxonomists gain a more complete picture of the organisms they have chosen to study and identify the gaps in knowledge about them that they hope to fill.

The **materials and methods** phase of a taxonomic study has two major foci: procurement of specimens and decisions as to the exact information that is to be recorded for each specimen or taxon. The data phase includes gathering activities, that is, measuring and recording the character values for the specimens along with preliminary analysis of these data. Distribution of the states of each potential character must be assessed and correlations that are of taxonomic value discovered.

The major primary **result** of a classical taxonomic study is the classification which demonstrates the collective pattern in the entire set of characters by displaying it in the traditional hierarchy of categories.

The **conclusion** to a monographic study is a descriptive monograph that is published. The classification, names, descriptions, and keys contained in it must all be prepared with great care since the monograph will be expected to serve as the definitive work on the group for some years to come.

Introduction: Objectives and Literature Survey

Definition of the Problem

In general, taxonomic problems can be divided into two kinds: cases where new data conflicts with the previous classification and cases where the previous classification has left a difficult pattern unresolved. New data may come from new kinds of comparative observations, such as chemical analysis of seeds, or it may be conventional data that has just become available, as in the case of specimens from the Amazon Forest and Planalto of Brazil. Unresolved difficulties in the variation pattern may be attributable to lack of time or to lack of material in previous surveys. Sometimes the use of new kinds of data or new ways of analyzing patterns allows a fresh look at problems previously thought to be intractable. We list some examples of monographic revisions that fit into one or more of these categories.

Example 1

The conference on the Solanaceae in Birmingham, England in July 1976 was held because of awareness that the time was ripe for a reassessment of the whole family. Much new comparative data, particularly from phytochemical and cytological surveys, and knowledge of much new material from Central and South America, meant that disagreements with the older classifications of Solanaceae were common. A new analysis of the pattern followed by rearrangement of the genera and subfamilies was required.

Example 2

In the 1960s Polhill started work on a monograph of African *Crotalaria* (rattlepods). Although Baker had catalogued many species in 1914, he had left the complicated pattern of interrelationships unresolved by opting for a crude, artificial partitioning. As more material became available, particularly from East Africa, and new species were described, as in the notes by Milne-Redhead in 1961, it became clear that the species showed a variety of resemblances which needed to be studied to clarify the picture. Polhill devoted much time and energy to surveying the materials and gave detailed attention to variations, some of which had escaped Baker's notice. In 1968 Polhill's work culminated in a new classification of the 434 African species; he divided them into ten sections. The monograph was devoted primarily to African species and neglected the 200 or so other species—a limitation imposed by time. However, work proceeds on the Indian species (by Rao) and the American species (by Windler), and the extent to which Polhill's classification will need modification in order to incorporate these remains to be seen.

Example 3

In 1978 one of us (DJR) started a survey of the wild species of *Vitis,* the grape genus. The taxonomy of the group is, of course, of interest to commercial grape breeders. The problem was one of an unresolved and difficult variation pattern. A detailed study involving more morphological characters and more thorough sampling of localities permitted clearer recognition of associations of characters and natural groupings, although problems of some intergradation among taxa remained.

Literature Survey

The literature survey in a taxonomic study often begins with the examination of various bibliographic indexes. These may include *Biological Abstracts,* which covers the current literature of all biological disciplines and is published in many installments throughout the year by BioSciences Information Service (BIOSIS); the Bibliography of Agriculture, published by the U.S. Department of Agriculture; the Commonwealth Agricultural Bureau Abstracting Service; and the Kew Record, which is an annual compendium of names and publications worldwide. All of these are now available as online computer databases. Also there are current specialized journals of the various groups of organisms, such as journals of bacteriology, of mammals, of birds, of ferns or mosses or whatever, that must be searched.

Because of the complexity of the literature, and the number of organisms that are known, taxonomists make considerable use of nomenclatural indexes. These are compiled by organizations that have taken on the responsibility of keeping the published record current by producing periodic indexes to the nomenclature. In botanical taxonomy for higher plants, two indexes, which list published taxa by family and genus, are important. The older is *Index Kewensis,* originated with funds donated by Charles Darwin and published under the auspices of the Royal Botanic Gardens Kew, in England. It lists all published names of seed plants under the appropriate higher taxon, arranged alphabetically by genera, and gives for each of them a bibliographic citation where the species name was first published. Generally, the major geographic area in which the taxon occurs is given. *Index Kewensis* is updated about every five years, and each new, updated issue of the Index contains 4000 to 6000 names that have been published since the previous issue.

The second index important for higher plant taxonomists is the Gray Card Index of Harvard University. This lists only plant taxa found in the Western Hemisphere. It includes the ferns as well as the seed plants, whereas the *Index Kewensis* lists only seed plants (Gymnosperms and Angiosperms). Supplemental cards are sent to subscribers of the Gray Index quarterly and the recipients add them to their card alphabetically by binomial name. The card format of this index has the advantage that all new, supplementary data can be added in alphabetical sequence. In *Index Kewensis,* where original and supplemental issues are separate bound volumes, there is no easy way to determine whether a new name has been added to a genus short of searching all of the separate volumes.

After indexes, monographs are the most valuable published resource. In addition to the specialized ones for lower taxa like genera, there are more extensive encyclopedic series of monographs. Two of the most complete treatments for higher plants ever published are Engler's *Das Pflanzenreich* (under new editorship from 1900 on) and *Die naturlichen Pflanzenfamilien* published by Engler and Prantl (1887–1915). Another outstanding reference is the three volume set *Genera Plantarum* by G. Bentham and J. D. Hooker (from 1862 to 1883). This monumental study, which describes genera of seed plants, differs from many such compendia in that the descriptions give particular emphasis to information coming directly from examination of the plants themselves rather than from previously published sources.

Animal taxonomists generally rely upon the *Zoological Record,* published by BIOSIS U.K. under the supervision of the Zoological Society of London. This outstanding record of the literature of animal taxonomy was started in 1864, is published every year in twenty sections, and for records from 1978 is available as an online database. The sections, as of 1978, deal with the various groups of animals as follows: (1) Comprehensive Zoology, (2) Protozoa, (3) Porifera, (4) Coelenterata, (5) Echinodermata, (6) Vermes (in three subsections), (7) Brachiopoda, (8) Bryozoa, (9) Mollusca, (10) Crustacea, (11) Trilobita, (12) Arachnida, (13) Insecta, (14) Protochordata, (15) Pisces, (16) Amphibia, (17) Reptilia, (18) Aves, (19) Mammalia, (20) List of New Generic and Subgeneric Names. Proper use of the *Zoological Record* requires considerable skill because of the very large number of names and worldwide bibliographic references. Mayr's (1969) discussion and description of the *Zoological Record* should be consulted for details.

There are a large number of abstracting services for major groups of animals. In malacology, for example, the scientist goes directly from the section on mollusca in the *Zoological Record* to the *Malacological Review,* which is a serial publication published twice a year at the University of Michigan. The *Malacological Review* publishes facsimilies of the tables of contents of all current serial publication on molluscs—nearly thirty journals from all over the world in 1982. The *Review* also contains a systematic index of papers and books on molluscs as well as original research reports.

Of course, the same places that have the specimens, the museums and herbaria, are frequently the most logical places to find specialized literature, because the publications are essential to the work of the museum. For example, two of the largest and most complete libraries of botany in the United States are found associated with two large, international herbaria— i.e., the New York Botanical Garden and the Missouri Botanical Garden. In the U.K. unrivaled taxonomic libraries are found at the Royal Botanic Gardens Kew and the British Museum (Natural History) in London.

Obscure publications are a major source of problems in literature searches. The earliest scientific journals or periodicals often had very limited circulations, so that papers published in them went unnoticed by taxonomists for many years. Indeed, many local natural history journals publish an occasional name even today.

When all of the pertinent literature has been found, taxonomists use it in two ways. The first use of the literature is to ascertain the variation in the group described by previous workers. In this way all taxa thought by others to be related, and the concepts underlying the previous classifications, will be considered. It may also prove useful later if the locations of specimens, particularly of rare or remote species, as well as of so-called type specimens, are noted during the initial literature survey. The second use for the literature search is to establish the correct nomenclature for the taxa as delimited at the beginning of the monographic study.

Materials and Methods:
Specimens, Data and Taxonomic Characters

Finding Specimens

Once a problem has been defined and the literature survey completed, the next step is to get living or preserved specimens in order to study the variation pattern. The literature survey will have indicated geographic areas or environmental habitats where living specimens may be found. Also from the literature, the worker may know where useful specimens are housed in the worldwide network of museums or herbaria. For many groups of organisms there are published guides to the collections, which can be of assistance in the search for suitable preserved materials for study. Addresses of herbaria are listed in the *Index Herbariorum*.

Bona fide investigators can arrange to borrow materials from herbaria for study. For a research student or junior staff member, loans are usually arranged through the professor or director of the institution, since there must be assurances that the specimens will be handled with care and returned in good condition within a reasonable time. The researcher is likely to be better received at the lending institution if he requests a reasonable amount of material—one does not ask for all of the specimens for a particular taxon or for representative specimens, since such requests would be exceedingly time-consuming for the curator. After initial analysis of the borrowed material, the worker may discover that it does not provide an adequate sample. Then, more specimens will have to be borrowed or plans made for collecting more materials from the field.

Gathering Data

The problem that one perceives often dictates the type of data that one will use. For instance, if a classification based on morphological observations is questioned because striking chemical resemblances are found between species that have been placed in different taxa, then clearly the new survey should include both morphological observations and the appropriate type of chemical data. Adding further surveys, possibly of cytological preparations, or of other classes of chemical substances, or of more detailed microscopic work on surfaces by use of a scanning electron microscope might also be considered. The purpose of adding these would be to discover yet more patterns in the data that might help to clarify the picture. The additional surveys could prove less useful if the species are either invariant or extremely variable for the characters observed. They can prove troublesome if some pattern is revealed, but this pattern does not agree with any of the original patterns. The most robust natural classifications are those having many types of data in which the variation pattern in one type is reinforced by the variation pattern in the others.

In Angiosperms and Gymnosperms the traditional data used for classifying have come from external morphology, and there are reasons why it is important to include a morphological survey in almost every taxonomic revision. For most groups morphological data are cheap and easy to obtain by comparison with, say, chemical data. Further, in most cases identifications will have to be carefully confirmed using morphological characters. Finally, morphological descriptions are a necessity for others to use in identification work.

There has been much interest recently in a variety of new data sources although few complete surveys on authoritatively identified materials have been carried out using these kinds of data. If a survey of pollen surfaces, anatomy, micromorphology, macromolecules or micromolecules, or cytology is contemplated, then it is a good idea to make a pilot survey first. If clear differences are observed in the pilot survey, then the taxonomist has evidence that the data type may prove taxonomically useful and is worth collecting. One is often influenced in the choice of the type of data collected either by one's own skills in the different techniques, or by those of one's supervisors or by the equipment that is available. It is no longer possible for every individual taxonomist to master all of the techniques that may be deemed useful. We shall not devote space here to the details of techniques related to the various data sources for plant taxonomy. However, at the end of the chapter there is a list of references pertinent to each of them.

Choosing Characters

The data phase is the part of a standard taxonomic analysis which, although carried out in all studies, is least well understood and about which there is much disagreement. The basic question is: how do taxonomists seek and recognize the variation pattern necessary for making a natural classification? We shall consider first taxonomic characters, which are the individual elements of variation observed, surveyed, and compared among the plants or animals; and then, in the Results section, the groupings—the associations of specimens or taxa formed as the classification takes shape either as abstractions or as physical piles of specimens. However, to a working taxonomist the analysis and refinement of taxonomic characters often goes hand in hand with considerations about potential groupings in relation to the overall pattern.

We define a **character** as a single basis for comparison among homologous parts within a given set of organisms. By "single basis for comparison" we mean a single or smallest comparable element such as a single structure or quality. Such a character is clearly a relative basis for comparisons that is peculiar to a given set of organisms or taxa. Were we to enlarge or reduce the set of organisms we are studying, characters useful for comparisons might change. If all the organisms we are trying to classify have hair, then the mere presence of hair is of no value as a taxonomic character in this study. However, if several specimens have straight, others wavy, some tightly curly hair and still others have had hair but are now bald, then we have a useful character Degree of Hair Curling with several states.

"Homologous parts" refers to the biologist's notion of **homology,** the idea that like must be compared with like, which we shall develop further below. Let us take as an example the degree of dissection of the leaves in legumes (members of the family Leguminosae or Fabaceae), as some of these should be familiar to botanists on all continents and will be found in nearly all illustrated flower or plants books. Fig. 3.1 illustrates some of the forms found in some legumes. In a study of Vicieae, the vetches and peas, we find only structures shown in *d, e,* and *i*—that is, leaves with no terminal leaflet (paripinnate) but with a varying number of pairs of lateral leaflets from zero or one to many. Here we assume that the structure in form *i* is homologous with a leaf despite the complete absence of leaflets! (The pair of structures at the leaf base are stipules.) If we look at the tribe Genisteae—the lupin(e)s, brooms, gorses, and laburnums—we find forms *a, b, c,* and *g*—simple, palmate, unifoliolate, and trifoliolate. In the Genisteae we could simply refer loosely to the number of leaflets but the case of *Crotalaria* reminds us that a simple leaf (form *a*) is not homologous with a leaf of only one leaflet as seen in form *c*. We see from the Genisteae and *Crotalaria* that the

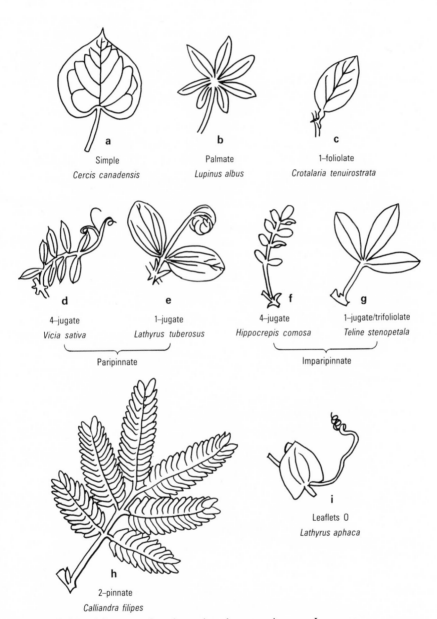

Figure 3.1 Leaf forms used to determine characters in some Legumes.

same structures (here form *a*) can be involved in slightly different comparisons in different groups. In studying the whole family Leguminosae a number of characters are needed to encompass the various comparisons that might be made—whether the leaves are compound or not, the order of pinnate branching, the number of leaflet pairs, the presence of terminal leaflets, and so on. Also note that when a number of characters is used to compare a complex of features in this way, the exact choice of characters is a subjective matter of perception.

Now that we have the idea of each character being a comparison of structural, qualitative, or quantitative differences, we can think of these differences being surveyed for sample specimens of plants to be classified. We may find, as in the leaf characters depicted, that the selected differences vary on some occasions quite distinctly between species but not within them (thus all leaves on all plants of *Lathyrus nissolia* are without leaflets and all leaves on all plants of *Vicia unijuga* are unijugate—that is, with only one pair of leaflets). In other cases the characters can vary from plant to plant within a species or, as in several species in the Genisteae, between leaves on different parts of the same plant. Thus, we see that each character has its own particular distribution pattern within a plant, within taxa under study and between taxa under study.

Taxonomists sift and select among large numbers of characters to find the so-called good characters. The minimum requirement is to determine whether the character qualifies as a taxonomic character at all. For this its distribution must be relatively invariant within each of the taxa or specimens to be classified, but show clear variation between some but not all of them. Thus, a taxonomic character allows us to perceive a pattern of resemblances and dissimilarities among the objects being classified. Of course in a biological classification differences revealed must be heritable differences, not short-lived environmental responses. Such a character would be qualified to play a part in the recognition and synthesis of a natural variation pattern.

Some taxonomists would go further and say that "good" characters are better than others. They may be better for classification purposes—that is, better because of associations between their distribution patterns and those of other characters. Or they may be better for diagnostic purposes, by virtue of being wholly of one form within one of the groupings and wholly of another form in others. So, for instance, the palmate leaf (fig. 3.1, form *b*) is a good classification character in the Leguminosae. It is found in most of the 200 or so *Lupinus* (lupin or lupine) species and is well correlated with other characters of that genus. However, it is not a good diagnostic character; some *Lupinus* species from Brazil have simple leaves.

In surveying the overall variation pattern many taxonomists think of each classificatory character as indicating an element of the pattern by its distribution. If a number of such characters are surveyed, it is possible to build up an impression of the collective natural pattern and it is this that is used to form the classification. Sometimes extra weight is given to particularly good characters; that is, those that correlate well with other characters. The distribution pattern of such characters contributes relatively heavily to the assessment of the overall pattern. This is called *a posteriori* weighting, as it is based on the known distributions of characters and is thought by some authors (see Davis and Heywood 1963) to be at the heart of good taxonomic practice.

An example of *a posteriori* weighting is the extent to which many Angiosperm taxa are delimited on the basis of floral characters. There are some families in which vegetative characters play a part in delimitation—the Rubiaceae, Palmae, Labiatae, and Gramineae for instance—but in many cases it is in floral characters that we find strong associations of characters. These are the bases of the classification in, for example, the Ranunculaceae, the Leguminosae, and the Solanaceae. The Solanaceae are delimited on the basis of a short list of associated structural characters of the flowers and fruits. Vegetative characters, such as leaf shape, size, and form, are highly varied. The only useful nonfloral characters are the bicollateral vascular bundles in the stem with phloem on both sides of the xylem.

A similar effort is made to recognize a collective pattern made up of contributions from individual characters when chemical, cytological, micromorphological, or pollen characters are used. So, for instance, in the very large genus *Solanum* of the Solanaceae (1600 species including the potato *S. tuberosum* and the aubergine/eggplant *S. melongena*) we find patterns of occurrence of a variety of alkaloids being used to give a pattern employed in the subdivision of the genus into sections.

All of the characters mentioned so far are intrinsic characters, properties of the plants themselves. We may also wish to take into account extrinsic properties, properties of the plants' interactions with the physical or biotic environment. Habitat preferences, geographic distribution, pollination mechanisms, and pollinator, insect, or mycorrhizal associations may all yield extra sets of characters, which may or may not reinforce the patterns already observed in other types of characters.

The characters that make the clearest contribution to a pattern are those that show sharp discontinuities: qualitative or structural differences, such as leaf form (as seen in Fig. 3.1), and discontinuities in quantitative measures. More difficult to handle are the less clearcut contributions from characters describing a continuum or nodes in a continuum. If the continuum is in a visual character that is difficult to quantify, a useful taxonomic technique is to construct an illustrated scale by mounting side by side drawings, photographs, or plant fragments (assuming that the fragments are from plant specimens in one's own collection and not specimens borrowed from a herbarium!). One of us (DJR) made use of such a scale of hairiness conditions on leaf undersurfaces for comparing *Vitis* specimens. Fragments were mounted in a logical order and used as a scale to score the specimens.

A variety of simple biometric techniques have long been used to help in the analysis of continuous characters and to some extent they are forebears of the models and computer methods described later in this book. For a simple character we can draw a frequency curve or histogram, as will be described in chapter 4, and look for multiple peaks and discontinuities. When assessing several characters, we can draw two-dimensional scatter diagrams, as will be described in chapter 8, and search visually for clouds of points and discontinuities in the diagrams.

It is the distribution pattern of each character that is important. The way in which this pattern does or does not coincide with the pattern of other characters determines its contribution to the overall pattern and hence to the classification. These properties do not depend on the extent of biological or functional knowledge of the character or on its adaptive significance. Some authors have argued that taxonomists must explore the functional aspects of each character. We agree, but not because this confers any improved classificatory power. Rather it helps in the recognition of homologies. Thus, the classificatory value of the occurrence pattern of a secondary substance, such as an alkaloid, is not improved by exhaustive chemical identification of the alkaloid. But identifying it does ensure that the same substance is being compared in the different specimens—or in short, that we are dealing with a homologous character. Understanding functions can also be the source of new characters, often behavioral or physiological characters, related to what was originally a single morphological observation. We are thinking of a case where a solitary observation of ant cavities in *Cecropia* (the plant genus) turned into a whole suite of characters on ant associations, ant behavior, and minor structural variations associated with these.

A view to which we do not subscribe is that nonadaptive characters should be favored. In the first place the history of people recognizing nonadaptive characters is littered with failures. Just think of the present interest in plant-insect substances, substances that were dismissed as nonadaptive byproducts as recently as the 1960s. Secondly, our prime concern is with a phenetic classification and an information system; we are searching for a stable classification on which we can hang comparative data that interest others, rather than for an evolutionary hypothesis based on supposed relict characters.

Results: The Classification

Natural Pattern and Groupings

The principal result of the revision is a freshly constructed classification. How do the taxonomists produce this? We have discussed in the previous section how they may survey characters and how these provide the pieces which, when put together, make the natural pattern. The process of putting together can take several forms. It can be done in the taxonomist's head, on paper, or by physically arranging groups of specimens. It can be based on repeated surveys of characters or repeated surveys of specimens.

However it is done, the detection of groupings depends on what taxonomists call correlation of characters. In its most extreme form this means the complete association among states of two or more characters or, if groups contain continuous characters, associations among discrete ranges of the continuous scales. Thus, *Ulex europaeus,* the species of gorse that grows wild in western Europe and is introduced in California, is clearly distinguished from other *Ulex* species by the complete correlation of two characters—the bracteoles being 2 mm or more wide and the somatic chromosome number being 96. More frequently something slightly less than complete correlation is found. We have already given the example of palmate leaves in *Lupinus,* which are not completely associated with other *Lupinus* characteristics because of the few simple-leaved South American species. Similarly, associations that make up many of the families are less than complete. In European genera of the family Labiatae, the ovary with four nutlets is distinctively associated with bilabiate (two-lipped) gamopetalous (with fused petals) flowers and opposite and decussate branching on square stems. In fact, this correlation does not always hold; there are a number of genera of the Labiatae where the ovary structure is rather different.

Correlated characters in different combinations yield the various taxonomic groupings with which we are familiar. We can phrase this differently to emphasize the groupings. Thus, *Ulex europaeus, Lupinus,* or the Labiatae are all groupings in which a particular combination of characters is present in either all members or most of the members, at the same time that the combination is absent in other groups. The extent to which the combination in one taxon and that in the most similar taxon differ gives us the taxonomist's concept of discontinuity. If there are no intermediates, but rather an abrupt change from the combination in one group to a different combination in the second, then we have a clear discontinuity. Alternatively, and this is common especially when large assemblages of characters are used, a minority of species or specimens complicate matters by showing other, possibly intermediate, combinations of characters.

In practice, there is a great deal of variation in the levels of correlation or in the levels of discontinuity found in natural taxonomic patterns. In the clearcut cases, such as in the examples given here, there is no difficulty in recognizing groupings, and the problems are concerned with marginal forms and whether to include them in large taxa or to segregate them in small offshoot groups. But there are also many cases where the character correlation itself is poor and there is genuine difficulty in detecting the patterns. What does the taxonomist do to detect patterns in these cases where they are not immediately obvious?

Method 1. The Building Up Method
The taxonomist reads through the data or leafs through the specimens, building up a picture of the most memorable combinations of character conditions. One then postulates that materials with this combination constitute a taxon and goes through the materials separating out those that belong to the putative taxon from those that don't. One then repeats the process on the remaining items, looking for another memorable combination to use for separating items, then a third, a fourth, and so on. This iterative process continues until all of the materials have been placed.

Method 2. The Gestalt or Overview Method
The taxonomist tries to picture the whole set of combinations. Each recurring combination is then thought of as the characteristic of a possible taxon. Then, the taxonomist works through the materials, either assigning them all to possible taxa or finding that there are some left over at the end. The leftovers may prove to be a new combination, one the worker had failed to notice, or they may be intermediates or outliers of uncertain position.

Method 3. The Subdividing Method
The taxonomist successively subdivides the plants into smaller and smaller groups until the subgroupings become homogeneous or impossible to further subdivide. Trial subdivisions are sometimes tested by drawing up a key and checking that the specimens or taxa can be unequivocally assigned to one group or another.

The three methods described are rather crude examples of what is really a wide range of subjective techniques for forming tentative groups. Depending on the complexity of the pattern and the taxonomist's ability to analyze it, at this point, there may or may not be a pattern which extends to several strata; that is, a pattern with subgroups within the putative groups. In any case a hypothesis has been formulated and it will be subjected to further testing and refinement.

In the process of **refinement** the taxonomist surveys yet more material, either some that has been deliberately held back, or some that has been freshly collected or recently received on loan. The question is: does this subsequent material fall easily into the tentative pattern? If the new material fits, it helps confirm the pattern and the researcher can move on. If it doesn't, one must investigate the reasons. This may result in redefinition of the limits of a taxon either by lumping or by splitting proposed groups. The appearance of intermediates may cause one to lump two putative taxa together. Alternatively, a new distinct variant within one of the common character combinations may cause one to split a group into two subgroups or completely sunder it into two main groups. Lastly, there is the possibility that completely new groups, not represented in the original survey, may turn up. These must be added to the pattern.

It is this refinement process that is so important for giving the taxonomist confidence that the notion of the pattern has moved from a hastily concocted impression based on limited observations toward a stable pattern which is supported by repeated tests on additional specimens. Refinement, as defined here, is equivalent to the hypothesis testing that is done in other scientific disciplines.

Finally, taxonomists subject their definitions of the natural variation pattern to a number of biological tests. If the study is of populations within species, then they may search for direct or indirect evidence of breeding barriers between the species proposed. They will look to see whether the proposed species occupy different geographical areas, and search for some differentiation in the ecological niches occupied by species whose geographical ranges overlap. At higher levels, say of genera or tribes, they will check on whether the putative taxa have contiguous ranges or if they span several continents, and whether their ranges correspond to known phytogeographic patterns. They will look at the cytological data to determine whether matching chromosome numbers or polyploid series add confirmation to the major groupings. None of these tests infallibly provide strictly objective criteria that are requirements for support of the tentative taxa. However, positive responses to one or more of the tests can be used as corroborative evidence.

Determination of Rank

Once the natural variation pattern is available, the next decisions concern the ranks to be used in the published classification. Are the smallest groups in the pattern species, sections, or genera? These can be difficult decisions to make because there is no objective way of telling the rank of a taxon with the possible exception of the species. The species is the nearest thing we have to a real biological entity, although there is hardly universal agreement on the definition of species. As Wiley (1978) puts it, "In all probability more paper has been consumed on the question of the nature and definition of the species than any other subject in evolutionary or systematic biology."

Three major conceptions of species have been promulgated over the past 50 years. They all recognize that the species is somehow the smallest natural population unit and the corollary that species are the entities of evolution. The **morphological species concept,** proposed by Du Rietz in 1930 (as quoted in Lawrence 1951) stated that species are "the smallest natural populations permanently separated from each other by a distinct discontinuity in the series of biotypes." The mechanism for ensuring this permanent separation is an integral part of the **biological species concept** (Mayr 1963), which defines species as groups of interbreeding natural populations that are reproductively isolated from other such groups. Because reproductive isolation, leading to distinct gene pools, is the mechanism proposed for maintaining separation between evolving groups, its presence or absence is seen as a potentially operational definition of species. Problems arise with the definition, however—especially with respect to asexual and non-outbreeding organisms. A more comprehensive definition, of which the biological species concept is a special case, is the **evolutionary species concept** of Simpson (1961). As modified by Wiley (1978) it states that "a species is a single lineage of ancestral descendent populations of organisms which maintains its identity from other lineages and which has its own evolutionary tendencies and historical fate." A problem with this definition is that separate lineages operate without producing predictable degrees of phenetic change and the practicing taxonomist is left to judge whether or not the observed morphological and geographic differences provide adequate evidence for separate evolutionary lineages in any given case.

There is evidence of extreme subjectivity regarding decisions with respect to ranks above the species level. Anderson (1956) experimented directly by sending a set of *Uvularia* specimens with the labels covered over to a number of botanical taxonomists for classification. He found that nearly all of them had grouped the specimens in the same way—that is, they observed much the same natural pattern, but they differed markedly in the ranks they assigned to the groupings. The results illustrate that rank is a relative matter. Taxonomists either compare the scale of resemblances and discontinuities with that of taxa in neighboring groups (something that was not available in the *Uvularia* study) or use the known upper and lower limits to the study to suggest the categories available. This means, for example, that a taxonomist working on the species of a tribe knows that the various strata in his concept of the natural pattern must be assigned to section, subgenus, genus and subtribe.

Walters (1961) has ventured the opinion that the degree of dissection of the classification is related to how thoroughly the plants have been investigated and that comparisons of wild and cultivated plants and of tropical and north temperate plants show these effects particularly well. We can contrast the highly dissected family Gramineae, which has separate genera for wheat *(Triticum),* barley *(Hordeum),* and rye *(Secale)* with the economically unimportant Cyperaceae, where a very wide range of variation is encompassed in a few genera, one of which is *Carex,* the sedges. Walters suggests that the dissection of the family Umbelliferae (the parsley family) is similarly exaggerated, in comparison with its neighbors, because it is a well known temperate family, while neighboring families in the classification are less well known plants of the tropics. If Linnaeus had worked in tropical forests, he asks, would the family Umbelliferae have been but one large genus *Umbellifer* in the Araliaceae?

Lastly, one of us (FAB) organized a roundtable meeting for about 20 specialists in the tribe Genisteae in 1977. There had been much instability in the classification of the tribe, which has a particularly complicated natural pattern. The object of the meeting was to review recent progress and to see whether some agreement could be reached for presentation to the International Legume Conference scheduled for a year later. The interesting outcome was that a broad agreement on the variation pattern was reached (see diagrams in Bisby 1981) but there was no agreement on the taxonomic rank of the groupings agreed to. Indeed, it was clear that what were treated as section and subspecies in the brooms *(Cytisus* spp.) of Western Europe were treated as genera and species by Eastern Europeans.

We conclude then that determining taxonomic rank is one of the more subjective decisions in taxonomy and offer only the following broad guidelines.

1. Try to place taxa at ranks which give a comparability of scale of discontinuities similar to that in the neighboring taxa.

2. Use the genus for the principal level of visual recognition. We note that it is often the case that a reasonably knowledgeable biologist recognizes, for instance, oaks *(Quercus),* maples *(Acer),* or willows *(Salix)* at sight but must check in a book before being certain of the species of each. This gives a general indication of the level of specificity characteristic of taxa designated as genera.

3. Be conservative and change ranks only where absolutely necessary, since there is considerable subjectivity involved.

Conclusions: The Publications

In taxonomic studies where classifications result, the conclusions are embodied in a series of formal publications of the kind described in chapter 2. A classification generally has the following parts: designation of the taxa and their ranks in the hierarchy, names appropriate to each of the taxa, descriptions of each of the taxa recognized, an identification key, specimen and literature citations. Thus, even after the classification itself is made, there are many things left to do to prepare the publication. We describe them below under headings that relate to the major sections of many monographs or revisions.

Introduction. In the process of arriving at the classification, the taxonomist has been guided by a set of objectives. These will have been stated, tentatively, at the beginning of the study and altered, refined, and amplified in the course of the work. To introduce his work and justify his conclusions, the worker must now prepare a concise statement of these objectives. If the conclusion is that the present classification is not adequate and a new one is needed, then the reasons must be specified. If the objective is a more specialized one, like producing a cytological classification based on the construction of the chromosomes, or an ecological one based on characters related to the environment, then these objectives must be stated clearly. Another matter that the taxonomist may want to discuss early in the formal presentation of the work is the data used. If new kinds of data have been added, such as descriptors from scanning electron microscopy or chromatography or gel electrophoresis, there is need to explain why these new, more sophisticated and refined measures are required for working with the organisms under study.

Nomenclature. When taxonomists publish a classification, they must apply the rules of nomenclature (see chapter 2) properly. If a taxon has been discovered for the first time in a study, a new name must be given to it in accordance with the rules for the group—be it animal, plant, or bacteria. The rules require, among other things, that the name be correctly formed in Latin, **effectively published,** which means published in a journal normally used by taxonomists for this purpose, **validly published,** which means given a Latin description, and that a nomenclatural type be cited. If the name being published is the name of a species, this nomenclatural type must be a specimen belonging to the species; and the type specimen, bearing collector identification, must be deposited in a museum or herbarium designated for housing such materials. If it is a genus or higher category name that is being published, the type must be a taxon of the next lower rank. Thus, the type of the family Rosaceae is the genus *Rosa*.

If the taxon has already been named, the rules are equally specific and require that the earliest name that meets the above specification be applied. Any other names that may have been used for the same taxon become synonyms.

Descriptions. An important and time consuming part of preparing for publication is writing descriptions of the kind defined in chapter 2. These need to be carefully checked so that ranges of variation as well as the most common occurrences are recorded for the characters. Descriptive information common to all subgroups of one taxon, as say to all species in one genus, should be given just once in the generic description. The more familiar the taxonomist is with the variation pattern, or the more complex the pattern is, the more likely it is that the generalizations in descriptions of higher taxa will have to be hedged with comments or exceptions. Mentioning these is important though; they may well be gratefully received, perhaps years later, by readers struggling to make sense of exceptions they have noticed.

People who use descriptions are often disappointed by inadequate comparability among the descriptors. A frequent problem is lack of precision with respect to common characteristics. For example, if the taxonomist knows that all species of broom *(Cytisus)* have yellow flowers except for one *(C. multiflorus),* it is a mistake to write descriptor states such as "flowers mostly yellow" in the generic description and "flowers white" under the description *C. multiflorus.* It leaves the reader uncertain as to whether other species have been examined and, in each case, the flowers found to be yellow. Similarly, mentioning structures in one description and omitting them from another can leave it ambiguous as to whether the structure is absent, present, or not examined in the second one.

In the written description there follows, after the description of the taxon, details of its distribution, and following that documentation on herbarium specimens. Information such as the herbaria from which the specimens were borrowed, their type status, and allied information will give the future investigators needed information that is not found in any other place. If space allows, other coments on habitats, plant-insect associations, palatability, and economic significance printed after the descriptions will be welcomed by many readers.

Keys. An important task in the conclusion phase is that of preparing a key. The taxonomist will attempt to identify characters that are easily observed and diagnostic, in the sense that they discriminate one taxon from another in an unambiguous fashion. Such characters are suitable for key characters. The key makes the information on the taxa included in the classification accessible to the nonspecialist user. In fact, keying out is probably the aspect of taxonomy most widely recognized by biologists of other disciplines, and it may well be true—if unfortunate—that the key will get more general attention than any other result of the taxonomist's work.

  Typification. One of the means for standardizing nomenclature of plants and animals all over the world is typification. As mentioned above the International Rules of Nomenclature require that for each new species name proposed, the author must lodge a specimen, which represents that name, in a public systematic collection—i.e., in a herbarium or museum. We call this **type method** and the specimen the **type specimen** or nomenclatural type. This designation may be unfortunate in that it is easy to confuse the meaning of the word 'type' in this context with the idea that 'type' means typical of the species. The latter is not the case; 'type specimen' simply indicates that the specimen is a member of the taxon to which a particular name has been given.

  In general, specimens provide documentation. The taxonomist, who cites the specimens in a collection examined in preparing a monograph, permits future investigators to determine the accuracy of any statements or interpretations made in the monograph. Even the oldest specimen can give valuable information. Very few other disciplines can claim documentation that extends back in some cases more than two hundred years!

  Annotation. In botanical work where herbarium specimens have been used, another activity for the final phase of the taxonomic study is annotation. The word is used to mean the process by which each worker, who studies a collection, documents his work by placing on the specimen sheets his own interpretation and classification. This ensures that the next worker who sees the specimen will be aware of what a previous worker has done. Annotation is therefore an effective means for communicating results among specialists on a group.

Suggested Readings

General

Davis, P.H. and Heywood, V.H. 1963. *Principles of Angiosperm Taxonomy.* Edinburgh and London: Oliver and Boyd.

Mayr E. 1969. *Principles of Systematic Zoology*. New York: McGraw Hill.

Stace, C.A. 1980. *Plant Taxonomy and Biosystematics*. London: Arnold.

Discussion

Burtt, B.L. 1964. Angiosperm taxonomy in practice. In V.H. Heywood and J. McNeill eds. *Phenetic and Phylogenetic Classification*. London: Systematics Association.

Disney, R.H.L. 1983. A synopsis of the taxonomist's tasks, with particular attention to phylogenetic cladism. *Field Studies* 5:841–865.

Walters, S.M. 1961. The shaping of Angiosperm taxonomy. *New Phytologist* 60:74–84.

Data Types for Angiosperms

Ferguson, A. 1980. *Biochemical Systematics and Evolution*. Glasgow: Blackie.

Harborne, J.B. and B.L. Turner 1984. *Plant Chemosystematics*. Orlando and London: Academic Press.

Heywood, V.H. ed 1971. *Scanning Electron Microscopy*. London and New York: Academic Press.

Leenhouts, P.W. 1968. A Guide to the Practice of Herbarium Taxonomy. *Regnum Vegetabile* 58.

Moore, P.D. and J.A. Webb 1978. *An Illustrated Guide to Pollen Analysis*. London: Hodder and Stroughton.

Moore, D.M. 1978. The Chromosomes and Plant Taxonomy. In H.E. Street ed. *Essays in Plant Taxonomy*, pp. 39–56. London and New York: Academic Press.

Part II

Theoretical Bases
for Taxonomic Work

Chapter 4

Structure
of Taxonomic Data

We have seen that investigations in taxonomy and systematics deal with large quantities of comparative data. We need to understand the formal structure of these data if we are to understand computer-assisted taxonomic methods. The specimens, species, or higher taxa that taxonomists study are, in general terms, **objects** or **items;** these two words will be used interchangeably. The attributes that they observe or measure for each object are the **descriptors** or **variables,** another pair of interchangeable terms. The **characters** used in taxonomic work, which we have referred to in previous chapters, are formulated from these descriptors or variables.

Tabular Data Structures

It is important to perceive the formal, structural distinctions among the items under study and the variables or descriptors used to characterize them. The items or objects are simply a series of comparable cases, that is, cases that are subject to description by a given set of variables. Variables, or descriptors, have two distinct aspects—their names and their states. The name of the variable defines a general property of the set of objects while the state indicates that particular description or measurement observed for an item. For example, for the qualitative descriptor Size and Location of Spotting in a study of *Salmo,* or trout (Legendre, Schreck and Behnke 1972) the three states of the descriptor with this name were "large, posterior," "medium, posterior and more anterior," and "fine and profuse, generalized." This example illustrates the relative terms, such as 'large', 'medium', and 'more anterior', which are commonly seen in taxonomic descriptors. Because these descriptors have meaning for the investigator, it can be seen that, even at the early stages of defining descriptors, the researcher already has a general overview of the variability within the data. An example of a quantitative descriptor is Number of Vertebrae, which may have state values ranging from 56 to 64. If the variable indicates a measurement like Length of the Pelvic Fin, then the measurement values themselves, such as 25.6, are the states. (The mathematically inclined will have recognized the term variable and have noticed that the distinction we have drawn between descriptor and descriptor state is directly analogous to that between variable and value in mathematics.)

We will now introduce a notation system. Objects or items will be designated by capital letters. When we wish to specify a representative object, we shall use **J**. For the general designation for the last object, we shall use **P**. (Note that this does not imply that there can be no more than the 16 objects represented by the letters **A** through **P** in the alphabet.) Descriptors or variables will be designated by lowercase, while the last descriptor will be **n** and the general variable **i**. Within each descriptor or variable the particular states or values will be designated by subscripts such that a_1 means the first state of variable **a**, a_2 means the second state, and so on, while b_1 means the first state of variable **b** and so on. The last subscript for any variable is m while the general one is k. Note that in some contexts we will subscript the variable letter name with a capital letter to indicate the state of the given variable for a particular object. Thus, i_J refers to the ith variable with respect to the Jth object.

We see now that a single observation on an item can be specified by designating a descriptor and a descriptor state. An observation can be unambiguously recorded as a pair like 'Plant Form, shrub,' which is in general, formal terms (De_i, ds_k) where De_i means the ith descriptor and ds_k is the kth state. In our terminology each observation, or piece of data entered, is subsumed in the designation i_k since this, in itself, indicates that the ith descriptor is observed in the kth state. Further, when many descriptors or variables are examined for an object, the object itself can be specified by a series of such recorded observations and the data on a single object is represented formally as:

$$[a_k, b_k, c_k \ldots \ldots, i_k, \ldots n_k]$$

We have said that work in taxonomy is based on comparative data. For data to be comparative, we require two things. First, the items in the study must be similar enough to allow us to compare like with like. This is what is meant in biological terminology by **homologous**. Second, each must be described, insofar as is practicable, by each one of a chosen set of variables. The usefulness of a collection of data for taxonomic purposes is degraded by missing data, although in many studies it can not be entirely avoided.

When a series of items is completely described for a set of variables, the data structure which results takes the form of a rectangular $n \times P$ table with n rows and P column headings for the specimens, species, samples, or whatever the items used. Having the descriptors as rows and the items as columns is a common, arbitrary choice in statistics, although recently in data-management parlance the opposite has become the usual convention.

The interior of such a table is composed of cells or spaces in which data entries are recorded either literally or in coded form. Each recording, or data point, is one state of one descriptor appropriate to one of the objects. Continuing our *Salmo* example, in Table 4.1, "bright orange" is the state of descriptor **b**, Color of the Top of the Dorsal Fin, observed for item **B**. Thus, "bright orange" is entered in the second space of the second row.

In formal notation a table like the one in table 4.1 is called a **matrix.** The cells containing the descriptor states are designated by symbols that relate them directly to the items and descriptors. For example, ds_{aA} refers to the descriptor state (ds) observed for object **A** on descriptor **a,** while ds_{ij} will take on one of the descriptor state values for a particular descriptor, the ith, that is, one of the m states in the descriptor state notation above.

There is no logical necessity that a cell contain only one entry.

Table 4.1 Tabular Data Structure for P Items and n Descriptors

Descriptors	A	B	C	. . . . .	J	. . . . .	P
				Items			
a Identification	101	102	103	. . . .	ds_{aJ}	. . . . .	ds_{aP}
b Color of the Top of the Dorsal Fin	cream	bright orange	cream and bright orange	. . . .	ds_{bJ}	. . . . .	ds_{bP}
c Rows of Scales Above Lateral Line	23	34	22	. . . .	ds_{cJ}	. . . . .	ds_{cP}
d Color of Cutthroat Marks	red	yellow	blank (no marks)	. . . .	ds_{dJ}	. . . . .	ds_{dP}
. .							
i	ds_{iA}	ds_{iB}	ds_{iC}	. . . .	ds_{iJ}	. . . . .	ds_{iP}
. .							
n	ds_{nA}	ds_{nB}	ds_{nC}	. . . .	ds_{nJ}	. . . . .	ds_{nP}

In fact, descriptors in the table may have either mutually exclusive (i.e., nonoverlapping) or overlapping states. In the former case only one entry is allowed per cell while in the latter a cell may contain more than one descriptor state. In the simple example of nonoverlapping descriptor states shown in Table 4.1 the dorsal fin color may be "cream" or "bright orange" but not both. If, in fact, the fish does have fins of both colors it is necessary to invent another descriptor state called "cream and bright orange." If overlapping descriptor states are allowed, then more than one descriptor state per item may be recorded. It is easy to see that the distinction is a formal one— the intrinsic reality of the data can be portrayed in either system.

There are two major aspects of taxonomic informational content for which the tabular data structure is less than ideal: within-taxon variation, and dependent characters. Both become more serious problems the more diverse the items are. Thus, large-scale floristic or faunal projects are affected more than monographic projects, especially those that are on taxa of species rank or below. There are ways—though sometimes clumsy ones— of structuring data so as to take variation and dependency into account and we will describe some of them here. The problems arise again with respect to appropriate structuring of databases for automated information systems to be considered in chapter 12.

Within-taxon variation is, of course, particularly important' when items are not individual specimens but populations or higher taxa—a frequent case in sytematics work. One way to deal with it is the overlapping tabular data structure, which allows for multiple entries per cell, possibly in coded form. Another obvious method is to enter ranges; these store the information but in a form that is textual rather than quantitative and therefore less well suited to analysis. Another way of dealing with ranges is to split the descriptors into a maximum indicator and a minimum one.

A different approach is to list each of the variable states for qualitative characters as a separate descriptor. The descriptor state of each one then becomes an indicator of the presence or absence of the attribute. When more than one of these descriptor states turned descriptor is marked present, the character is clearly a variable one.

Dependent characters are those that are applicable only if another character is present. Thus, our variable regarding cutthroat marks could have been structured into two descriptors: (1) Presence of Cutthroat Marks and (2) Color of Cutthroat Marks. Then, applicability of the second descriptor would have been dependent on the state of the first; if the first were marked "absent" then the second would be "nonapplicable." The information is recorded and stored in either case, but any direct use in biological comparisons of the term "nonapplicable" is, of course, not appropriate. Such descriptors must be restructured into comparable characters before use in analyses. If several levels of dependency are involved the effect of subsuming them into single characters, as in our Color of Cutthroat Marks variable in its original form, is to produce a complex character having many states. Such variables are awkward to use in retrieval queries. Further, the formal structure of the information they contain can influence automated taxonomic classification methods in special ways that will be discussed later.

Data Types

Descriptor states come in a variety of forms. We distinguish two broad categories of data—quantitative and qualitative—and a number of variants of each. Knowing the properties of the several data types is a matter of importance. Most methods of analysis and of information storage and retrieval are appropriate only for particular data types.

Quantitative data must be expressed in numbers; however, not all of the numbers used have the same properties. One type of quantitative data, which we shall refer to as **metric** because it is used for measurements, is expressed in real numbers—numbers that can take on any decimal value limited only by the resolution of the measuring instrument. For example, it is reasonable, from a mathematical standpoint, to express the weight of a pumpkin as 10 kg, 10.2 kg or 10.2473 kg; however, in practice there is a limit to the number of decimal places, which is set by the sensitivity of the scales used to weigh the pumpkin.

The other major type of quantitative data is termed **discontinuous.** The common case is that where integers are the only reasonable descriptor states, as in counts or sets of ranks. Again, the integers used do not always have the same properties. The **ordinal** property of relative position, without the **interval** property of absolute distance, is sufficient to express ranks. To illustrate, if plant **A** blooms before plant **B** and after plant **C,** we can rank them by assigning 2, 3, 1 respectively. The numbers express the relative blooming times of the three plants but do not provide information on the distance between **A**'s blooming time and **B**'s. Counts are an example of discontinuous numbers which have interval as well as ordinal properties. The absolute size of the interval between two leaves and six can be reliably discerned!

There are times in the preparation of character states for classifying when it is necessary to consider whether a particular set of discontinuous character states has simply ordinal or ordinal plus interval properties. A further question which arises in the latter case is whether the states are evenly spaced or not, that is **linear,** or **nonlinear.**

It is worth noting for those who will want to do statistical anal-

64 THEORETICAL BASES

yses of their data that there are special methods corresponding to each of the three quantitative data types. Metric data are used for statistical significance testing that is based on means and variances compared with normal curves. Enumerative statistics, designed for count data, provide significance tests comparable to those for metric data. This is not surprising, since count data differ from metric only in lack of continuity; corrections can be made to take this into account. The type of statistics known as nonparametric is appropriate to ranked data. Such data, having only ordinal properties, contain less information. Thus, it is more difficult to achieve significance with nonparametric statistics. However, they have an advantage in that they are not dependent on the normal distribution.

Qualitative data are those descriptor states in which names or sets of words are used to define a condition observed for an object. The states are discrete and there may be two or more of them so that the term **multistate** is often used. **Binary** descriptors are those that have only two states. There are two distinct types of binary descriptors: **symmetric** and **asymmetric.** Symmetric binary is simply a special case of multistate—the biological context being such that only two states are of interest. Asymmetric binary is that dichotomous data with descriptor states such as "present–absent," "yes–no," "0–1." Only two states are logical possibilities, and one of them is of different significance than the other. It is important to know which of the binary types one has, since appropriate analyses differ.

Qualitative descriptors are often represented by numbers. Numbers save space, are more convenient, and are suitable for computer input. Of course, such coding in no way changes qualitative data into quantitative data.

Qualitative data may in some cases be orderable and thus acquire a property usually associated with quantitative data. We shall call such data **ordered** as opposed to **simple.** In the example in table 4.2 the first descriptor has simple qualitative descriptor states while the second illustrates an ordered qualitative descriptor.

Several important points are illustrated in these two examples. In both cases the descriptor states are coded by the same set of symbols. In

Table 4.2 An Example of Ordered Data

Descriptor	Descriptor states		
Color of Cutthroat Marks	red	yellow	blank (no marks)
ds code:	1	2	3
Size and Location of Spotting	large, posterior	medium, posterior and anterior	fine, profuse
ds code:	1	2	3

the simple type the symbols are assigned arbitrarily while for the ordered one there is a logical assignment related to the order inherent in the biological content. Further, the ordered descriptor states may or may not be evenly spaced. In our example it probably makes biological sense to assume that there is more difference between 2 and 3 than between 1 and 2. We shall designate such cases as **nonlinear** and those in which the ordered values are set apart by equal intervals as **linear.** There are provisions in some computer programs which allow one to assign the values of intervals so that nonlinearity can be taken into account. Fig. 4.1 summarizes the data types and fig. 4.2 gives a botanical example of the various data types that were encountered in one taxonomic study.

The two examples of qualitative data types also illustrate that the appropriate form of analysis will depend on the meaning and not alone on the formal structure of the data. Multistate data of the ordered type, which can be assumed to be linear, can be handled like quantitative data. This is true because means of ordered descriptor states have a reasonable interpretation especially when there are three or more states. On the other hand averaging the numeric symbols in the case of the simple qualitative descriptor gives a meaningless result. For this reason such coded data should not be analyzed by most of the common statistical techniques, nor by certain classification procedures (see chapter 7).

There are ways to transform data types from one to another. For example, it is always possible to convert an $n \times P$ table, of the qualitative multistate type, into a table of binary data of the asymmetric type by listing all of the descriptor states as descriptors (the same procedure we suggested above as a means of dealing with variation within taxa). Two possible states will then exist for any given item—yes or no, often coded 1 or 0. In this case the data structure is still tabular, but its new form makes it suitable for

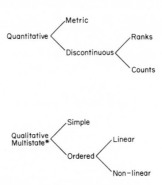

Figure 4.1 Summary of data types.

*Includes binary of the symmetric and asymmetric types.

Qualitative (double-state) characters
 (1)   (i)   *Calyx lobes* acute (1) or rounded (0).
 (2)   (ii) *Receptacle* hairy inside to the base (1) or not (0).
 (3)   (iii) *Ovary* bilocular (1) or unilocular (0).
 (4)   (iv) *Bracts* and *bracteoles* enclosing the flowers in small groups (1) or not (0).
 (5)   (v)   *Stamens* far exserted (1) or not (0).
 (6)   (vi) *Staminodes* united into a 'comb' (1) or not (0).
 (7)   (vii) *Ovary* terminal at the mouth of the receptacle (1) or not (0).
 (8)   (viii) *Stipules* enlarged (1) or not (0).
 (9)   (ix) *Two glands* present at the base of the lamina or on the petiole (1) or not (0).
 (10) (x)   *Lenticels* prominent on young *flowering* stem (1) or not (0).
 (11) (xi) *Stamens* united to form a single ligule (1) or not (0).

Qualitative (multi-state) characters
 (12) *Receptacle shape*: elongate, symmetrical, hollow (1), elongate symmetrical, solid (2), ventricose (3), saccate (4).
 (13) *Fertile stamens* occupying the whole (1), two-thirds (2), or less than a half (3), of the perimeter.
 (14) *Leaf undersurface* with stomatal cavities (1), softly and densely hairy (2), glabrous (3), with stiff but not dense hairs (4).
 (15) *Bracts and bracteoles* with many sessile or stalked glands (1), two sessile basal glands (2), or none (3).
 (16) *Epicarp* verrucose (1), smooth (or with a few hairs) (2), with a dense rusty tomentum (3), a dense covering of crustaceous warts (4).
 (17) *Endocarp* smooth (1), very rough and fibrous (2), hard and roughish (3).
 (18) *Seedling-escape* by basal stoppers or plates (1), a single line of weakness (2), three or more lines of weakness (3), no special mechanism (4).
 (19) *Inflorescence* a panicle (1), corymbose panicle (2), elongated raceme (3), short subcorymbose raceme (4).

Quantitative characters
 (20) *Stamen number*.
 (21) *Flower size* measured in millimetres from articulation to apex of calyx lobe.

Figure 4.2 Example of data types in a botanical study. *(From Prance, Rogers, and White 1969:1211 and 1212)*

certain computer methods of association and information storage and retrieval.

In taxonomic work there are many significant qualitative characters. They are often in addition to measurement characters, so that the result is a mixed data set. This presents no problem if all of the qualitative characters are ordered, since these can be effectively transformed so as to be consistent with quantitative characters. Another example of such a transformation of qualitative to quantitative data is that of asymmetric binary in which the non-zero responses may be counted and the data analyzed by enumerative methods.

In the large data sets used in systematic work in biology, there are commonly many characters which must remain qualitative since they are of the simple type. In this case, the methods used are often those suitable for qualitative data, and the quantitative data may have to be transformed unless a program is available having mixed data capacity, such as Gower's (1971) mixed data coefficient. If necessary, it is always possible to convert quantitative data to qualitative, of the multistate ordered type, by partitioning the measurements into a sequence of classes. The process loses some of the precision of the information, but it does allow comparability of quantitative and qualitative data for purposes of analysis. (Note that there is no problem with simple storing of mixed data on most information management systems.)

    The major disadvantage of converting quantitative data so that it is, structurally, qualitative multistate is that the techniques for analyzing such data are not as highly developed and well known as those for metric data. This has resulted in undue emphasis on measurement variables in work involving numeric taxonomy. However, there are some good methods for classification using qualitative and mixed data which are described in chapter 7.

Formulating Descriptor States

Scaling

Scaling and partitioning are two standard means of processing descriptor states so that they are suitable for comparison in classification programs. **Scaling** is applied to metric data recorded in the original measurement units. It is obvious that a descriptor whose states are in millimeters can not be compared directly with one having states in kilograms. Relative magnitudes of diverse metric characters can be compared if the original values are scaled. There are a number of rules for scaling; two of the more important ones are standardizing and scaling by range.

    The first step in **standardizing** is to calculate the mean and standard deviation (i.e., the average deviation from the mean) of the values for a given variable. Second, the mean is subtracted from the value for each object and the result divided by the standard deviation. The numbers produced are the standardized values, often called the z scores, of the original data points. Note that the transformed data set always has a mean of 0 and a standard deviation of 1. If the data are normally distributed, the effect of scaling is to allow more than 99% of the values in any data set to be expressed by some number between $+3$ and -3. Even for non-normal data these limits will include almost 90% of the data values. Scaling by standardization has inconvenient features as well: it gives negative values, has no absolute maximum or minimum, and is best suited to normally distributed values.

If the minimum value of the entire set is subtracted from the value for each object and the result divided by the range of all of the data (i.e., the largest value minus the smallest value), the outcome will be a positive number between 0 and 1. This set of values constitutes another scaling of the data, **scaling by range.** In this case the values are easier to manipulate than those from standardization. However, for probabilistic data the effect of sampling error is increased because the mean and variance are more reliable quantities than the range. There is always the possibility that the single points that represent maxima and minima of any data set are nonrepresentative.

Partitioning

As we discussed above there are a number of reasons why quantitative metric descriptors must sometimes be partitioned into qualitative, ordered descriptors. One method of partitioning is to divide the total range of values into a convenient number of classes by assigning class limits either arbitrarily or according to some biological expectation. For example, the *Salmo* character Rows of Scales above Lateral Line was categorized so that, for example, those having 25.0 to 32.9 rows were assigned to State 3, while those having 33.0 to 37.9 were assigned to State 4, and so on. Note the use of the decimals to avoid ambiguity with respect to assignment of measurements at the end points of the ranges.

The effect of partitioning by ranges is to produce fewer, cruder values with greater rounding error than was in the original data. Certain information is lost. However, this disadvantage may be minimal in that often unnecessary or even unreliable detail is eliminated. As mentioned above, partitioning is frequently necessitated by the requirement of a computer program for multistate data alone.

Another procedure for partitioning metric data is suggested by the intuitive methods frequently employed by traditional taxonomists. In this the first step is to make a **frequency distribution** of the values recorded for a descriptor. This means deciding upon a set of class intervals, counting the number of cases or objects that fit into each, then graphing this information. The **histogram** or frequency diagram that results is shown as the shaded area in fig. 4.3. By connecting the midpoints of the bars with lines, a **frequency polygon,** which is also shown in the figure, is formed.

This polygon is examined for break points—that is, abrupt changes

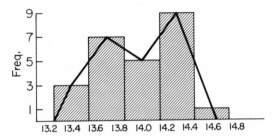

Figure 4.3 Histogram (shaded area) and frequency polygon (dark lines), which can be used to establish descriptor state ranges.

in direction or steepness of the line graph. In the bimodal distribution shown in fig. 4.3, there is such a break point, and one could divide the data into two groups using as criteria "less than 14.0" and "more than 14.0." Groupings or partitions that are defined according to such break points can be defended as being more directly related to the data itself than is the case with arbitrarily imposed intervals. In some cases it may even be reasonable to interpret them as natural classes signifying to some degree the genetic variation pattern within a group with respect to the given descriptor.

Nontabular Data Structures

There are a number of data structures, used in systematic biology, that are not $n \times P$ tables. They may contain any kind of data—quantitative or qualitative—but the data are arranged differently. An important case is that of **pairwise data.** When the data collected pertain to some relationship between members of a pair of items, the result is a $P \times P$ table in which the rows and columns are both lists of the same set of items. Pairwise data come from comparisons made in serology, hybridization experiments, DNA hybridizations, and other similar situations. To fill out the $P \times P$ table completely, observations must be made on all possible pairs. Since the comparison of an item with itself is usually set by definition, up to $P (P - 1)$ experimental observations are required. Such comparisons may or may not be symmetric; this is to say, the comparison **(A, B)** may or may not be the same as **(B, A).** If they are symmetric, the number of observations necessary is cut in half.

Fig. 4.4 portrays graphically the results of hybridization trials among *Geum* species. These data could be made into a $P \times P$ table in which,

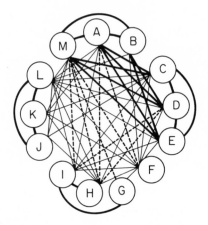

A - G. coccineum   H - G. laciniatum
B - G. silvaticum   I - G. canadense
C - G. urbanum   J - G. perincisum
D - G. hispidum   K - G. oregonense
E - G. molle   L - G. macrophyllum
F - G. aleppicum   M - G. rivale
G - G. boliviense

Figure 4.4 Diagram of pairwise, hybridization relations among species of *Geum*. Thick lines indicate fertile F_1 hybrids, dashed lines partially fertile hybrids and thin lines sterile hybrids. *(From Briggs and Walters 1984, fig. 10.7, 194)*

for instance, the entry corresponding to species *G. laciniatum* (H) in the rows and *G. urbanum* (C) in the columns is "sterile hybrid." Notice that in this example not all of the possible comparisons have been made; as a result, the table would have some missing data.

It is difficult to incorporate original data of the pairwise type into data management systems because they are structurally incompatible with tabular data. However, pairwise data can provide usable input for classification analysis by computer because they are in a form appropriate for the second step of the classification process. As we shall describe in chapter 7 the first step in two-step classification is to summarize the data in all of the descriptors by calculating overall similarity measures—i.e., single figures representing degrees of similarity among all possible pairs of objects. Originally pairwise data can easily be used in place of this $P \times P$ table.

A different kind of relationship among objects is seen in the case of pedigree data. Keeping track of parentage relationships for a large number of items presents especially difficult problems. The general structure of the data is that of a tree. However, the actual tree for a set of data represents only a particular subset of all the possible trees that could occur in accounting for relationships among the objects. Further, pedigrees are time dependent, in that a beginning point must be chosen. There are certain conventions that have been established by plant and animal breeders along with the standard methods of path analysis for summarizing degrees of relationship. Notice that in the case of the latter the data again revert to the form of a $P \times P$ table.

There are also instances where square data matrices of the $n \times n$ type result from pairwise comparisons of the variables. The variance-covariance or correlation matrices commonly used in multivariate statistics are good examples.

Now that readers have been introduced to the standard tabular data of taxonomy, to its notation and its exceptions, they should be able to move on to the more interesting models and analytical algorithms described in the next chapter.

NOTE: For **suggested readings** for this chapter, please see chapter 5.

Chapter 5

Models and Algorithms

N ow that we recognize the tabular structure of the data, the task is to organize and arrange the specimens or taxa on the basis of the descriptors, so that they form a hierarchical pattern: that nested-boxes arrangement of groups and subgroups which we have already shown to be the regular form for the classification of a group of organisms and which, in turn, is the basis for other taxonomic products. Strictly hierarchic structure of this type is deeply ingrained in the thinking of biologists. Relationships among life-forms have long been diagrammed as treelike structures, in which repeated branchings represent smaller and smaller subdivisions within the hierarchy. Evolution of increasingly diverse life forms over time is also easily represented by tree structures and this has long been seen by some biologists as validation for the naturalness of the hierarchic system of classification of organisms. The hierarchical assumption has led to endless argument as to whether a ''good'' classification either can or should approximate the actual evolutionary descent of a group. Be that as it may, hierarchic classifications *are* convenient, efficient means for organizing much biological information; there can be little argument about that.

There are biological purposes for which the nonoverlapping, multilevel hierarchy is not ideal. For example, an ecological study may require organization of species and data about their environmental requirements into a single-level partitioning. Or it may be advantageous to order the partitions into a graded series and perhaps to relax the strictly nonoverlapping requirement in order to show up transitional organisms. In taxonomy, hierarchic information structure is almost all-pervasive, so we shall concentrate on the general and theoretical aspects of the taxonomic hierarchy.

One of the first things that became apparent when the computer's potential for assisting the taxonomist was noted was that taxonomists would have to examine their thinking rigorously. Their semi-intuitive decision-making processes had to be made explicit and then translated into step-by-step procedures or algorithms suitable for computer programming. It sounded straightforward enough, but one of us (DJR), an experienced taxonomist who subjected himself to intense grilling by a computer mathematician on ''how he did what he knew how to do'' (Rogers and Tanimoto 1960), found that the task was by no means an easy one!

The results of such efforts have, however, yielded important insights into the theory of taxonomy, in addition to the computer programs produced for assisting the practical work of taxonomy. We shall define here three important models, which reflect important aspects of taxonomic thinking, and characterize some major algorithms that are used to produce hierarchic organization of information from the primary source, which is the tabular data structure.

A **model** is a generalized, simplified representation of the complex system or set of data under study. A particular model is selected because there is a correspondence between its properties and some—seldom if ever all—of the properties of the problem. A model is often used to suggest testable hypotheses. Thus, models are useful analogies, which may come from a subject area quite different from that of the problem at hand. This is particularly true of mathematical models, whose strength for biological problems lies in their ability to provide a system whose logical properties have been derived and proven. The biological system, to the extent that it corresponds to the model, can expectedly be described by these, and only these, logical properties. This system can serve to direct biological thinking and hypothesis testing and often to reduce the number of options to be considered. An illustration of this was seen in the early 1960s when a major problem in biology was to crack the genetic code. Rosen (1959) pointed out that the algebraic group properties of the problem of coding four DNA generators into about twenty amino acids predicted the one gene–one protein concept.

Algorithms are simply procedures for problem solving. The set of steps one learns in school for doing long division is a good example of an algorithm. Consideration of algorithms has become prominent in these days of programming computers, since the machines are not given to intuitive leaps of understanding, and must instead by directed as to how to proceed at each step along the way. It is often possible to apply the same algorithm to quite different models. We shall see that this has happened in the course of applying clustering procedures to taxonomic data and that there has been confusion between models and algorithms at times.

Models for Taxonomic Hierarchy

At the risk of oversimplification we will present the mathematical bases for taxonomic hierarchies in terms of three major models. The models are not entirely separable; there are important relationships among them. However, since the three are based in different types of mathematics, it seems useful to begin by dealing with them separately and pointing out their distinctive properties and purposes. After that we shall examine their interrelationships and then the more general aspects of the actual algorithms by which a tabular data structure may be transformed into a hierarchy via any of the models.

The three models which we shall consider are: the **geometric** model, which is the model for multivariate statistics generally; the **graph theoretic** model, which comes from the graph and set theory of finite mathematics; and the **information theory** model, which is a model based on information measures of diversity. Each has properties which make it suitable for particular kinds of data or problems. For example, the graph and information theoretic models are well suited to qualitative data, including simple multistate, while the geometric model fits metric or ordered multistate data.

It is interesting to note that to some extent the three models parallel the historical development of computer-assisted taxonomy in that there were three groups, each of which contributed a major method. The numerical taxonomy or NT group, associated with Sokal and Sneath, employed multivariate statistical methods and the geometric model (see Sneath and Sokal 1973). The taximetrics group, originated by Rogers and Estabrook, utilized graph theory models. The more ecologically oriented Southampton-Canberra group of Williams, Lance, Lambert, and Watson, whose work is generally described in Clifford and Stephenson (1975), contributed important models for classification from information theory.

The Geometric Model

The geometric model utilizes a representation of items or objects in space. Thus, if there are two variables—that is, two measures of descriptor values recorded for each item—then the space is two-dimensional and the geometric representation is a simple scatter diagram. The axes, at right angles, stand for the variables, and the position of the item is determined according to the scale of possible values for that variable which is marked off on each of the axes. The resulting data points represent the items. It is easy to extend such visualization to a three-dimensional or solid model, but beyond three dimensions there is no obvious physical interpretation. However, in theory a point can be plotted in relation to any number, *n,* dimensions simultaneously and the algebraic properties of the extended representation remain perfectly valid. We speak of such an *n*-dimensional extension of the simple scatter diagram, in which the dimensions are variables and the points items, as the **multidimensional geometric model** or **hyperspace model.**

Most people find it easy to think of plotting points in space against axes when there are numerical values for each of the variables. Therefore taxonomic methods that presume to use a geometric model are most intuitive when there is metric or at least ordered multistate data in the data matrix. There is a further reason that metric data are especially well suited to multivariate geometric models. In taxonomic work we often compare variables which have unlike units of measure. It is easier to interpret a plot of, say, leaf length in centimeters against tree height in meters or seed protein in milligrams if the data have been transformed so that they are unitless. The usual way to accomplish this is standardization after computation of the means and standard deviations of the variables (chapter 4). Such calculations require quantitative data.

Once the points have been plotted in multidimensional space the resulting distribution of points may be analyzed so as to reveal the hierarchic structure in the distribution of the objects themselves, which is the matter of concern to the taxonomist. Taxonomic objects are not seen as *existing* in hyperspace. They merely correspond formally to the points in the model so that the logical and mathematical properties of the model can be utilized in organizing information about them. One such property of the geometric model is that two items having all of the same values on the variables will occupy the same point in the space. The converse is also true: if the values are different on any of the variables, then the objects will occur at different points. From this primary assumption (axiom) we may define several taxonomic features by analogy with geometric properties of the model.

Dissimilarity between taxonomic objects is represented by distances between points in the geometric model. Taxonomic groups, such as species, are represented by clusters of points having short multidimensional distances within them—that is, having homogeneity in taxonomic terms. In addition there may be discontinuities, seen as gaps in the hyperspace, separating clusters one from another thus making clusters correspond to taxonomic groups. To apply the geometric model by analogy to the taxonomic hierarchy one must think in terms of successive fusions of nearby clusters. There are several ways to envision the determination of distances between clusters and they will have different implications for a taxonomist. For example, take the simple geometric idea of representing a cluster by its centroid (in which pairs of points are replaced by their midpoints, which are then used in succeeding calculations of distances). In a taxonomic analysis, this concept means representing a taxon by a derived point which probably corresponds to no actual taxon. Another problem in comparing clusters arises from the different degrees of dispersion within clusters, which correspond, biologically, to different degrees of variability within taxa.

We have so far considered only what are known as existence properties of the geometric hyperspace model; that is, we have claimed that it is reasonable to assume that clusters exist and that distances can be measured, but we have not presented any means for actually doing the job of determining cluster membership or measuring the distances. The latter is the province of algorithms. The analogy with the case of long division is that we have now shown that there is an answer for 1924 divided by 1257 but we have not shown how to find it.

We have seen above that there is no physical representation for a space of more than three dimensions. This means that we cannot actually visualize clusters. Even though we can to a degree reason by analogy from the two- and three-dimensional cases, we need to establish special algorithms for making calculations and determinations in hyperspace. First we describe various means for measuring distances in the geometric hyperspace model.

Direct Distance Measures (Dissimilarities)

If we envision a two-dimensional scatter plot, it is easy to conceive of measuring the distance between two points. If the axes are at right angles to each other, that is, **orthogonal,** there is a simple way of finding the distance between two points by using the familiar Pythagorean rule for right triangles: The square of the hypotenuse is equal to the sum of the squares of the two sides.

In Fig. 5.1 we see that the distance between objects **A** and **B**, designated as Di_{AB}, can be found as a function of the values for each of the objects on each of the variables **a** and **b**. These coordinates of the point are represented as (a_A, b_A) and (a_B, b_B) read as particular values of the descriptor state that is observed for object **A** on variable **a** and so on. Thus, the square of the distance along the variable **a** is found as $(a_A - a_B)^2$ which corresponds to one of the sides squared in the Pythagorean rule while the other side squared is $(b_A - b_B)^2$. Note that since the distances are squared the direction of the subtraction does not matter. The general formula for distance between two objects, **A** and **B**, in two dimensions is, then:

$$Di_{AB} = \sqrt{(a_A - a_B)^2 + (b_A - b_B)^2}$$

This measure is termed the **Euclidean distance** between two points and measures the shortest distance between two points in the plane.

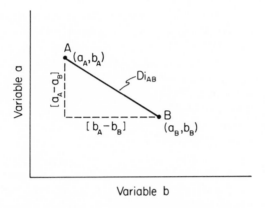

Variable b

Figure 5.1 Euclidean distance in two-dimensional space.

The Euclidean distance measure has the fortunate property of being simply extensible to three or more dimensions. As an example, we show below a small portion of the data table from Rohlf's study of mosquito species (1962). The data are measurements on the three characters Wing Length in mm, Length of Tarsomere I of Hind Leg in ocular units, and Cerci Length in ocular units, for the two species *Aedes aurifer* and *A. excruscians*. Calculation of the distance between these two species using these three variables is shown in fig. 5.2.

Similarly, the rule for Euclidean distance can be extended to any

		Species Number	
		5	12
Adult	#28	5.4	6.1
	#36	7.0	8.0
Characters	#45	2.3	2.2

$$Di^2 = (5.4 - 6.1)^2 + (7.0 - 8.0)^2 + (2.3 - 2.2)^2 =$$

$$-.7^2 + -1^2 + .1^2 = 1.50; \quad Di = 1.225$$

Figure 5.2 Example of the calculation of Euclidean distance between two species.

number of dimensions (variables). Thus, we have a simple measure of distance which can be applied to any and all possible pairs of points in n-dimensional space.

Although Euclidean distance is most commonly used, it is by no means the only measure of distance. Besides the angular measure that will be discussed below, there are other formulations of the linear distances between objects or groups. One important one is the **Manhattan** or **city block distance** measure, which is a simple summation of the absolute differences (that is, the magnitudes of the differences without regard to their sign) between two objects on each of the variables taken in turn. When this sum is divided by the number of variables to get an average, the Mean Character Difference, MCD, of Cain and Harrison (1960) results.

Distance measures of all kinds, as well as similarity measures (see below), can be evaluated with respect to certain of their general properties. Three are of particular significance:

(1) The distance between two objects calculated by the distance rule or function under consideration is always a positive value greater than zero. Or, to put it differently, a distance of zero implies that two objects are the same.

(2) The measure is symmetrical in the sense that the distance between objects **A** and **B** is the same as that between **B** and **A**.

(3) When three objects are considered, one can always designate **A, B,** and **C** such that the distance from **A** to **C** is less than or equal to the sum of the distance from **A** to **B** plus that from **B** to **C**. This property is known as the triangle inequality, since it is always true for any triple of points that can be arranged physically as a triangle.

If all three of these properties hold true for a certain distance or similarity measure, it is said to be a **metric measure.** (The term is not to be confused with metric data described in chapter 4 or even with the metric system.)

Not all distance functions in common use are fully metric. Even though theoretical considerations would lead one to prefer that the metric quality be a minimum requirement for such measures, practical considerations sometimes dictate the use of **semimetrics** or **quasimetrics.** The first type of nonmetric violates the first condition above in that $Di_{AB} = 0$ does not necessarily mean that **A** is the same item as **B.** We shall see an example of this below in our discussion of angular measure. The quasimetric is a type in which the distance function does not fulfill the triangle inequality requirement. Euclidean distance squared is an example.

When a new distance or similarity measure is proposed, it is of interest to determine whether or not it is metric. In some cases mathematical proof is difficult, but distance measures of the type called the **Minkowski**

metrics have been shown to be fully metric. All members of this group are formed by finding the absolute distances between two objects on a variable and raising that number to some power, r. This process is repeated for each variable, using the same r, and the resulting values are summed; the rth root of this sum is the distance. Notice that the Euclidean distance and Manhattan block measures discussed above are both Minkowski metrics, for $r = 2$ and $r = 1$ respectively.

Angular distance measure

Another approach to measuring distance between items plotted in hyperspace is to draw lines from each item to the origin of the axes of the hyperspace and then measure the angle between this pair of lines. The larger the angle between the lines the greater the distance between the pair of points and the more dissimilar the items are presumed to be. The geometry is shown, in the simple two-dimensional case again, in fig. 5.3.

The distance, Di_{AB}, between two objects **A** and **B** is measured by the absolute size of the angle between them. The lines drawn from the origin to the points **A** and **B** are known as vectors or rays, and the angle between them can be found by the methods of matrix algebra applied to standardized data. Though the size of the angle can be calculated it is obvious that it has certain limitations as a measure of distance. Figure 5.3 shows the case of distance from **A** to **C** which subtends the same angle as the distance from **A** to **B** even though the former would appear to be larger. Furthermore, two objects on any ray would have an angle of zero between them but would not necessarily be the same object. This property is interpreted favorably in some taxonomic situations in that it is thought to allow consid-

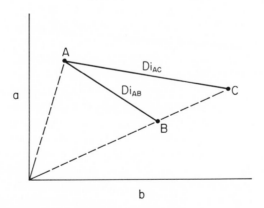

Figure 5.3 Angular distance measure in two-dimensional space. Vectors are shown as dotted lines, distances as solid lines.

eration of shape without the influence of size. However, it is clear that the measure is not a metric one.

Such determinations of angular measures between objects are termed correlation measures and are much used by the Sokal and Sneath Numerical Taxonomy school. Although the computer computation is done by matrix algebra manipulation, the result is the same as the ordinary statistical correlation coefficient. Statistical correlations, as explained below, are usually measures of the degree of association among variables. In this case, however, the determination of angular distance is equivalent to determination of correlation among objects. The procedure for calculating the Pearson's product-moment correlation coefficient, r, (see any standard statistics text) is applied to the columns instead of the rows of the matrix. This is hard to interpret. It means that one compares each of two objects' values on each variable in turn with the mean of its values on all variables; that is, the arithmetic average of such measures, for a single specimen, as leaf length, height, fruit size, etc. Further, these values are not usually standardized since variable (row) standardization is generally performed rather than object (column) standardization or both. This problem with respect to standardization raises further questions as to the accuracy of the correlations computed by matrix methods as measures of the cosines of the angles between objects.

Correlations Among Variables

We have been discussing associations among objects in taxonomic studies as being analogous to clusters in hyperspace. It is also possible to consider relationships among the variables themselves on the basis of the geometric model. Although the relative emphasis on variable relationships is less in taxonomy than in some other areas, there are some important applications of variable analysis in taxomonic studies. Furthermore, correlations among variables in hyperspace must be considered in making proper interpretations of clusters in the multidimensional geometric model.

The degree of association among the variables in a study is termed **correlation,** and can be described in terms of predictability: if two variables are correlated, one can, by knowing the positions of an object with respect to one of the variables, make a better than chance prediction of its position on the other. There are a number of standard statistical methods for calculating correlation. A common one is Pearson's product-moment correlation coefficient, r, mentioned above. In the case of r, or most other correlation coefficients, the result is a single number or coefficient whose values range from $+1$ to -1. A coefficient near 1 means that a high value on one variable is associated with a high one on the other while a negative coefficient, near -1, implies association of high values on one variable with low values on the other. Correlation coefficients near zero are low correlation values and mean that no regular prediction is possible. Variables whose correlation coefficients are near zero are said to be independent while those which are either significantly positively or significantly negatively correlated are dependent. Very high values of correlation, especially in biological studies where variability tends to be high, may mean that the variables are simply redundant: they are two measures of the same phenomenon. This is not to imply that redundant characters are never useful to taxonomists; they may or may not be.

Independent and dependent variables or dimensions can be represented differently in multidimensional space. Independent dimensions are orthogonal and thus represented by axes that are at right angles to each other. Dependent variables are pictured as being at other than right angles with the value of the correlation coefficient r equal to the cosine of the angle between the two variables. (In this book when we refer to the geometric model we are usually thinking of that multidimensional space in which all axes are mutually perpendicular.)

Taxonomic and Statistical Correlation

The meaning and use of correlated characters in taxonomy is an area of great complexity and a source of considerable disagreement. The difficulties are reflected in the arguments over character weighting that have raged since numerical techniques for classification were first proposed. Use of correlated characters is essentially equivalent to weighting certain characters. Many who use numerical techniques, wishing to be as objective as possible, view the use of characters known to be correlated as *a priori* weighting and therefore improper. On the other hand traditional taxonomists commonly claim that classification "requires correlated characters" (Anderson 1956).

When we examine what taxonomists mean by correlation among characters, we find that their use of the term and that of the statisticians differ in important ways. When speaking of correlation the taxonomist refers to a stable association, presumably genetically based, among certain character states. Statistical correlation also requires predictable association of values for any given pair of variables. However, the two concepts differ, particularly with respect to the role of discontinuities. To a taxonomist, correlation is worth noting only if it is associated with clusters. Statistical correlation is equally valid for data points that are spread out so as to indicate some continuous trend. In fact, measures of statistical correlation assume linear continuity in that an association of high values on one variable with high values on the other implies a similar association of low values. This is not true of taxonomic correlation.

Let us illustrate these points with some hypothetical examples. In fig. 5.4 the cases shown in both (a) and (b) would give a high degree of statistical correlation but only the situation in (a) would be considered useful

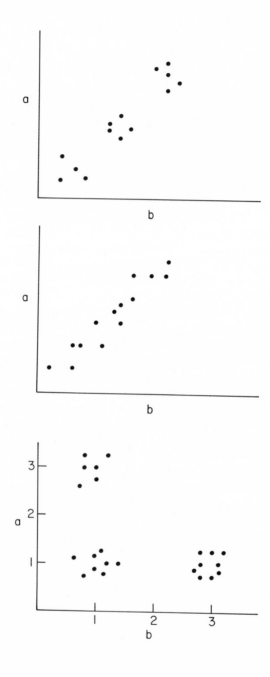

Figure 5.4 Taxonomic versus statistical correlation: (a, upper) taxonomically useful correlation; (b, middle) statistical correlation; (c, lower) taxonomic correlation without statistical correlation.

by the taxonomist. Note that the role of the two characters **a** and **b** taken together is to reinforce the division of the objects into groupings that would have occurred on either character singly. This is commonly the case when a taxonomist finds two characters usefully correlated.

Looking at the opposite case, in which there is taxonomic correlation without statistical correlation, we see in fig. 5.4(c) a situation in which three groups are separated along two dimensions. In this case the two variables are expressed individually; they do not act entirely in concert to make the divisions. Also, computation of statistical correlation yields relatively low correlation, since there is no pronounced linear trend. Still, the characters are useful taxonomically because they demonstrate a predictable association of states near 1 or 3 of variable **a** with states near 1 on **b**, along with an association between states near 1 on **a** and 3 on **b**. Taxonomic power is further enhanced by the fact that when the two characters are considered together they yield three clusters, whereas on the basis of either character alone there would have been only two.

In summary then, even when either fully metric or ordered multistate character values are used, taxonomically useful correlation is not completely predictable from statistical correlation. It is worth noting again that statistical correlation is not even a valid concept when using simple multistate characters, but taxonomically significant associations of character states are often found with such characters. In fact, it is the taxonomically correlated discrete states of either dichotomous or disordered multistate characters that are the usual criteria for distinguishing higher level taxonomic groups like order or family. Distinctions based on metric characters are more typical at lower levels. Many of these quantitative characters are normally distributed rather than multimodal, which tends to weaken their discriminatory capacity and accounts for some of the instability of groupings at the lower taxonomic levels. The difficulty is not entirely avoidable in that one may have no other characters to use. However, it points out the need for techniques that can make use of mixed data—that is, qualitative in combination with quantitative—and for means for analyzing the predictive capacities of qualitative multistate characters. Both of these topics will be discussed below.

Principal Components Analysis and Ordination

Principal Components Analysis (PCA) is frequently used in taxonomy for analyzing relationships among variables. The process of PCA results in a transformation of the original variables into a new set of axes, now called **components.** The reason for the transformation is that these components have useful properties. First, they are derived in such a way that they are all mutually orthogonal—that is, they are independent. Second, the PCA transformation is designed to arrange the components in order. The first

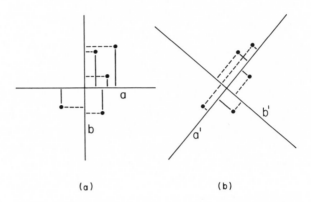

(a)   (b)

Figure 5.5 Rotation of axes in PCA to secure line of best fit among objects.

component accounts for as much as possible of the variability in the original data, the second for the next largest portion of the variability, and so on.

Geometrically the Principal Components can be visualized as the new set of axes, all at right angles to each other, which most closely approximates the scatter of the data points (objects) plotted in the original hyperspace. The geometric interpretation is drawn in fig. 5.5 for the two-dimensional case.

The original representation of the objects plotted against the two variables, **a** and **b**, is shown in fig. 5.5(a). In fig. 5.5(b) the pair of axes has been rotated in such a way as to best fit the data points with respect to both variables simultaneously. This has the effect of markedly reducing the scatter of the points about the axis marked **a'**. We can quantify this reduction by evaluating the distances between the points and the axes. Notice that the sum of the squares of the distances indicated by the solid lines is much less in (b) than in (a). We interpret this by saying that a large portion of the total variance is accounted for by the first component, which is represented geometrically by **a'**. Another geometric interpretation of the Principal Components is to regard them as the axes, arranged in order of decreasing length, of an ellipsoid (a multidimensional ellipse) whose center is the origin positioned at the multiple mean.

The components can be uniquely specified in relation to the original variables. Algebraically each component is representable as a linear combination of the original variables in which each of these variables is weighted, that is, multiplied by an appropriate weighting value, w. These products are then summed for each component. As an algebriac expression, then, the first component is:

$$C_I = w_{Ia}\, \mathbf{a} + w_{Ib}\mathbf{b} + w_{Ic}\, \mathbf{c} + \ldots w_{Ii}\, \mathbf{i} + \ldots + w_{In}\, \mathbf{n}$$

while the second is

$$C_{II} = w_{IIa}\,\mathbf{a} + w_{IIb}\,\mathbf{b} + w_{IIc}\,\mathbf{c} + \ldots w_{IIi}\,\mathbf{i} + \ldots + w_{IIn}\,\mathbf{n}$$

and so on.

The weights, often termed **loadings,** are unique for each variable on each component. They too have a geometric interpreetation. They are functions of the angles between the variables and the components. One can think of the component as a line in hyperspace whose optimal position results from particular rotations of the axes defining the dimensions (variables). The cosines of these angles of rotation are the values which must be calculated to find the loadings.

Reasonable as the idea may sound, it is hardly obvious how one would calculate such values—especially for angles in a theoretical space having more than three dimensions! The problem is not, in fact, so difficult. This is because it is a form of a problem that is central to linear algebra known as the eigenvector or latent vectors problem. Fortunately, the user of PCA does not need to know how to calculate eigenvalues and eigenvectors. The computer can be used to do that; the taxonomist need only supply a square, symmetric matrix and know how to interpret the outcome.

The $n \times P$ data table is the basis for a square table that is the input used directly for computing the principal components. This square table is an $n \times n$ matrix, the so-called variance-covariance matrix computed on variable (row) standardized data. It is simply a table of the correlations, r, among all possible pairs of variables. Since it is a square and symmetric matrix, it can be subjected to eigenvalue factorization by a suitable computer program. From this one can find **eigenvectors,** which provide the weights or loadings needed. Since, as shown above, these required weightings are equivalent to the set of angles of rotation, it makes sense that one transforms a matrix of correlations, which represent angle relationships between variables as defined in the last section.

The output from the computer includes eigenvectors for the components. These are sometimes used to determine which, if any, variables load heavily on particular components. If several load together on one or more components, it may suggest important relationships among these variables. The analysis of eigenvector values for this purpose is known as **reification** and is highly developed in factor analysis of which, in some senses, PCA is a type.

The computer output for PCA also usually contains so-called **eigenvalues** from which are computed the cumulative proportions of the total multiple variance that is accounted for by each component. Since the PCA procedures assign the majority of the variance to the first few components, we have a means for in effect reducing the dimensionality of the data. For example, if the first three components contain 80 percent of the variability on ten variables, one may feel justified in substituting these three components for the ten variables and thus simplify subsequent data analysis.

One significant use made of reducing the number of effective variables is in **ordination**—plotting the objects in reduced space by using two, or sometimes three, of the components as dimensions. To make this plot one needs to transform the original scores for each of the objects to coordinates on the components that will be used as axes for the plots. This process makes use of the eigenvectors or loadings again. One simply incorporates the eigenvector values into the linear combination formulas shown above along with the values for a single object on each of the variables, **a** through **n,** doing this for the first component and then for the second.

An example of an ordination comes from a study of disease diagnoses in relation to standard blood variables (Abbott and Mitton 1980). The first component has the eigenvector values shown in column 3 of fig. 5.6. The values correspond to the standard blood variables, measured on an autoanalyzer, named in column 2. The measures on these blood variables (standardized) for patient number 638 are given in column 1.

Multiplying the figures in column 1 by those in column 3 for each blood variable and summing the results gives the position of that patient with respect to Principal Component 1. Similarly, eigenvectors for Component 2 are in column 4. Multiplying these also by the measures in column 1 and summing the results gives the position for the same patient with respect to Principal Component 2. Transformed values may be found

Column 1 Standardized values for patient 638	Column 2 Blood variables	Column 3 Component 1	Column 4 Component 2
.338	Serum urea nitrogen	.498	−.546
− .517	Blood glucose	.341	−.005
−2.808	Carbon dioxide	−.418	.498
1.278	Chloride	−.405	−.630
.518	Sodium	−.658	−.448
−1.156	Potassium	−.186	−.583
.336	Calcium	−.659	.121
1.048	Alkaline phosphatase	.107	.427
− .582	Total bilirubin	−.006	.671
1.107	Albumin	−.772	−.070
− .605	Globulin	.484	.291
− .395	SGOT	−.241	.458
.567	Hematocrit	−.706	.315

Figure 5.6 Standardized values for a patient (Col. 1) and loadings on components (Col. 3 and Col. 4) resulting from a PCA on standard blood variables (Col. 2). *(From Abbott and Mitton 1980. Table 3)*

for every patient on each of the components and most of the PCA computer programs will put out these figures on any selected number of components. This was done in our example for the first two components and the objects plotted as shown in fig. 5.7(a).

Scatter diagrams, which result from plotting the objects after ordination, may be examined for trends or clusters. Three clusters are indicated in our example by the shading in the figure. In this case the clusters correspond fairly closely to the previously diagnosed diseases—hepatitis, cirrhosis, and chronic renal disease—indicated by the different plotting symbols.

Ordination is frequently used in taxonomic studies, especially for indicating major groups. An example involving the classification of birds is seen in the work of Schnell (1970). There are some cautions which must be observed in interpreting plots of points representing objects in principal components space. Usually the original variables are not all mutually independent as are the components. (Indeed, if they were there would be nothing gained by doing a PCA, since the object distribution along the components and the original variables would be equivalent.) Thus, distances among objects in principal components space should not be interpreted as if they were simple transformations of the distances in the space of the original variables.

Another useful technique for identifying groups in a set of objects is to use their transformed (principal components) scores as the input to a hierarchical clustering program using an algorithm of the kind that will be presented in the last section of this chapter and discussed more fully in chapter 7. The result is a dendrogram of the sort shown in fig. 5.7(b). In this truncated dendrogram we show only the major branches which correspond to the clusters in the plotted ordination. While the correspondence is obviously predictable, the dendrogram from cluster analysis has some advantages as a means for displaying groups. It has the potential for revealing more of the detail in the relationships between objects. Further, transformed scores on all of the components can be used as input. Thus, less information is lost than in the two-dimensional ordination.

Discriminant Function and Generalized Distance

For taxonomic applications we are particularly interested in analyses of the multidimensional geometric model that indicate clusters or, to take the opposite point of view, emphasize the gaps between clusters. One method for doing this is called Discriminant Function Analysis (DFA). In it we take a different approach to multivariate analyses, in that we do not extract groups either of objects or of variables, as is done in cluster analyses or ordinations. Instead, DFA is a method for testing preclassified groups.

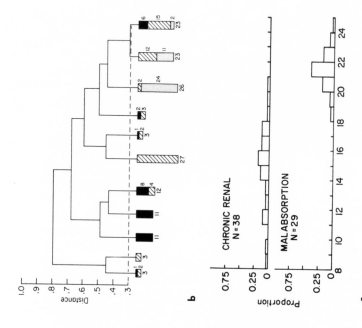

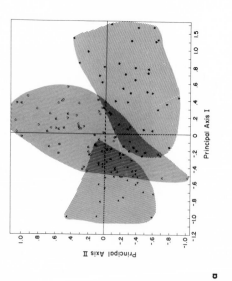

Figure 5.7 Diagrams of objects grouped according to their transformed scores following PCA: (a) an ordination or scatter plot in reduced dimensions (triangles represent previously diagnosed chronic renal disease, filled circles hepatitis and open circles cirrhosis); (b) dendrogram showing groupings obtained by cluster analysis (filled bars represent hepatitis; crosshatched bars cirrhosis, and dotted bars chronic renal disease); (c) frequency diagram from discriminant function analysis of data for chronic renal disease and malabsorption. *(From Abbott and Mitton 1980. Figs. 1, 2, and 4 respectively).*

Once a function that discriminates successfully among these groups has been calculated, that function can then be used in a predictive capacity to assign unknown objects to appropriate categories.

DFA is designed to maximize the distance between just two groups. Each group is characterized by the same set of multiple variables. In addition, objects in each group have been pre-assigned to one of two categories, on the basis of a qualitative descriptor (not one of the multiple variables, which are generally metric). To illustrate with the medical diagnosis example we used above: Patients in the study had all been diagnosed to a disease group. For each there were also data on the 13 variables resulting from a battery of standard blood tests. Consider the two disease groups Chronic Renal Disease and Malabsorption. The question is whether the blood variables can be used simultaneously to distinguish between the two diseases.

To investigate this we again form a weighted combination of the variables, as in PCA. However, this time just one linear combination is formed and the weights assigned to the variables are devised so as to maximize the distance between the multiple means of the variables for the objects in each of the groups. The discriminant function thus specifies an axis through the means of the two clusters in the multidimensional geometric model that maximizes the space between them. Once this axis is designated by the linear combination, individuals' scores can be transformed to correspond to it in a fashion exactly analogous to the PCA transformation. These scores are usually presented in a frequency diagram as shown in fig. 5.7(c). The horizontal axes represent transformed values for individuals within each of the diagnostic groups. Numbers of cases for the various values are shown vertically by the bars of the histogram. Thus the efficacy of the discriminant function is shown by the degree of overlap of the two distributions.

Now that we have considered the interpretation of the discriminant function, let us briefly summarize its construction (Marriott 1974). It is clear that one seeks to maximize the distance between a pair of points, the multivariate means of each cluster, and one might think a simple Euclidean distance measure would suffice. In practice the problem is complicated by the fact that variables are usually correlated rather than independent. (See section above on correlation among variables.) To take account of this one calculates the variances and covariances among all possible pairs of variables. The multivariate difference of means between the two groups (a vector) is multiplied by the inverted form of the resulting matrix of correlation coefficients. The effect of this is to correct for the variable correlations by dividing by a measure of them. The result is a single, linear combination of the variables called the discriminant function.

The same idea of compensating for intercorrelations among the variables can be applied to any distance measures in the multidimensional geometric model. Such measures between two points, be they original data points or means of clusters, are known as generalized distances or Mahalanobis D^2. They are calculated by squaring the distance vectors between the two points, using appropriate matrix operations, and then multiplying by the inverted matrix of correlations, as in DFA.

In taxonomic practice one is often interested in discriminating among more than two groups. This can be accomplished by a multiple discriminant function known as canonical variates analysis (or Multivariate Analysis of Variance). Its use in taxonomy will be described in chapter 7.

Principal Coordinates Analysis (PCO)

There is an important interrelationship between PCA and PCO which manifests itself in what Gower (1966) terms the duality between analyses based on variables and analyses based on objects arranged in that variable space. PCA is primarily a variable analysis. The variables are restructured in the process to the form of orthogonal (mutually perpendicular) axes known as components. As we have shown, the coordinates of objects, represented as points in the component space, may now be found and the objects plotted in the transformed space.

We also know that distances between objects can be measured by the Euclidean distance or other measures using the original variable values. The set of such distance or similarity measures resulting from all possible pairwise comparisons of objects is the first step in an object analysis and forms a $P \times P$ table of similarities that can be analyzed. Gower (1966) showed that in some cases one could begin either with variable analysis, PCA, or with object analysis using a table of similarities or distances, and arrive at the same distribution of points in the geometric space of the principal components. (The latter is not, of course, generally the same as the space of the original variables, nor can the coordinates of the points on the original variables be recovered.)

The process of computing the coordinates of objects along the principal components from dissimilarities or distances, without ever knowing the original variable values or their variance-covariance matrix, is known as Principal Coordinates Analysis (PCO). When the PCO is applied to Euclidean distances or other Minkowski metrics, the same coordinates that would result from a PCA are derived. The PCO methods can also be applied with exact results to similarity measures, including a simple matching coefficient for binary data that has been converted to dissimilarity by $(1 - S_{IJ})^{1/2}$. Other similarity measures may be used to approximate the principal components solution although computational problems arise unless the values are all positive. The PCO technique is useful to taxonomists. Applications will be discussed in chapter 7 and examples given.

Nonmetric Multidimensional Scaling (MDS)

Another approach to plotting multivariate data in reduced numbers of dimensions, usually two, is multidimensional scaling. The general idea of the method is the following. If one has a set of distances between all possible pairs in a group of several objects, there is likely to be sufficient information to define their configuration in space (except for mirror images and rotations). It is easy to think of this when the distances are on a plane—it is like making a map on the basis of a set of distances between several cities. The requirements imposed by the relative distances between the cities will define a particular map (which lacks only compass points) although determining that map, given only the distances, may not be easy. In MDS this concept of plotting relative distances on a plane is applied to distances that are multidimensional. Again the problem is to ascertain a unique solution. There are computer programs for doing this that proceed by repeatedly improving on an original approximation to the solution.

A form of this kind of scaling analysis that is useful for taxonomists is the nonmetric multidimensional scaling (MDS) first derived by Shepard and Kruskal in the 1960s (Shepard 1980). In this case, the original distance matrix does not have to be metric as we have assumed in the example above. Any similarity or distance matrix can be used. The original similarities or distances are considered to be nonlinearly related to the two-dimensional representation in that the only critical requirement is that monotonicity or rank-order be preserved. This is to say that ordinal properties are sufficient and that, for example, similarities expressed merely as ranks may be mapped into a two-dimensional representation of the items.

The object in MDS is to produce the most distortion-free reproduction possible in two dimensions; that is, the one which best preserves the rank order of magnitudes of the similarities or distances. One begins with a configuration of a few points chosen at random or with an initial configuration of the points that is thought to be a reasonable approximation. Points are added or their positions in the two-dimensional configuration adjusted in small steps. The results at each iteration are tested for degree of distortion using the following measure of what is termed stress, S:

$$S = \left\{ \sum_{J,K} \frac{(Di_{JK} - \widehat{Di}_{JK})^2}{\sum_{J,K} Di_{jk}^2} \right\}^{1/2}$$

The Di_{JK}'s are Euclidean distances among all possible pairs of objects and Di_{JK} is the distance to which Di_{JK} must be lowered or raised to take its correct position in the rank order, so that

$$\sum_{J,K} (Di_{JK} - \widehat{Di_{JK}})^2$$

gives the sum of the squared deviations from an ideal fit. This is then expressed as a fraction of

$$\sum_{J,K} Di_{JK}{}^2,$$

the total squared distance in the two-dimensional space. The process of adjusting and testing for stress continues until convergence is achieved at the point where further adjustments fail to improve the correspondence significantly.

When we began this section on geometric models we suggested the most easily visualized interpretation—that of plotting data points (objects) in a space in which the variables serve as axes. This simple approach would appear to limit the usefulness of the geometric model to metric variables in that the operation of plotting against axes presupposes quantitative data. In fact the geometric model is applicable to nonmetric data as well in many cases. Extension of the usefulness of the geometric model is underlain by Gower's (1966) demonstration of an equivalence between the information in a table of distances (provided that the distances have the properties of metric measures discussed above) and that in a data matrix. This means that any data set that can be expressed as a set of distances (or similarities) is representable in the geometric hyperspace model.

This is an important result for taxonomic analysis. As in the example given above of determining a map given the distances between cities rather than their map coordinates, the taxonomist may now properly envision objects in a multidimensional space even though his information about them is based on relative similarities among them. PCO and MDS are clear examples of how one can use a geometric model even though the distances used are calculated from data that are not entirely metric. Similarly distance or similarity measures based on a variety of data types can be used in a cluster analysis and the results thought of as groupings arranged in a geometric hyperspace.

Graph and Set Theoretic Models

The set theory of mathematics, which has become familiar to many via the recent "new" mathematics, is a rigorous, systematic statement of concepts that are fundamental to mathematical logic despite their apparent simplicity. A **set** is defined as a collection of like elements or items. In addition there is associated with it a rule or criterion by which one may determine whether or not a prospective element belongs to the set. Sets may be overlapping or mutually exclusive; that is, an element may belong to more than one set simultaneously or it may belong to one and only one set. Sets may be divided into subsets. If a set is divided into two or more mutually exclusive and exhaustive subsets (that is, every point is assigned to only one subset), then the set is said to be partitioned. Application of a series of criteria in this way will result in nested partitions and thereby a strictly hierarchic structure.

It is easy to recognize the model for taxonomic classification in these set-theoretic terms. The specimens or taxa to be classified can be thought of as a set that will be sequentially partitioned. We have said that sets are associated with rules of assignment. In taxonomic work the characters provide the capacity to determine whether a specimen does or does not belong to a species or a species to a family. To be more precise it is the presence or absence of a selected state of a character, or a combination of such states, that is used as the criterion. This is straightforward enough in the case of simple multistate characters. To use a quantitative character in such a rule of assignment one must group its values into ranges. For example, in the definition of *Acer glabrum* (fig. 2.2) one criterion for assignment to the species is trunk diameter between 0.5 and 5 inches.

There is a further consideration that applies to taxonomic work. As noted above a hierarchic classification must result in a series of partitions. Obviously the two or more states of a single character can be used as criterion for only one partitioning. Taxonomic data always include multiple variables; how to use them all is the question. There are two major approaches. The first is to use a sequence of variables to make the succession of subdivisions required. This results in a series of **monothetic** groups: each group is defined by a single character state or a combination of characters (connected by the Boolean 'and') that is present in all the members of one group and absent in others. The method requires choosing the order of application of the characters to be used as criteria. Usually these decisions depend on the biological judgment of the taxonomist as to the appropriateness of different characters at the various levels. In actual practice most taxonomic assignments are made on the basis of several taxonomically correlated characters. For example, the presence of hair, nursing young, and four limbs are highly associated character states which are used monothetically, but act together to reinforce assignment to the Class Mammalia.

In the **polythetic** method of classification the objects are grouped on the basis of shared character states over several characters. Objects in a group will be alike with respect to a number of character states but no particular ones are either necessary or sufficient to define group membership. Clearly, objects that share like states of many characters will be more simi-

lar than those which share only a few. The degree of similarity between individual items is calculated by considering all characters and is expressed
quantitatively in a **similarity measure** or **coefficient.**

There are a large number of such similarity measures that have
been used in taxonomic work; their properties will be discussed in chapter
7. One example is the simple matching coefficient, in which the number of
character states for which two objects match is divided by the total number
of characters considered. It should be noted here that all of the distance measures discussed above are a type of similarity measure. Actually, they are
measures of dissimilarity in that an increase in similarity is indicated by a
decrease in distance. Similarity (or distance) measures are calculated for all
possible pairs of objects. The result is a $P \times P$ matrix. Similarities are usually defined in ways that produce diagonally symmetric matrices. The values
along the diagonal are all measures of maximal similarity—either ones for
similarities or zeros for dissimilarities. It is this matrix of pairwise measures
of the degrees of similarity among the objects that forms the basis for the
graph theoretic model of hierarchic structure.

Graph theory provides a more refined model for the process of
partitioning, which we have been describing in general set theoretic terms.
In graph theory the elements in a set *and* the relationships among them are
depicted in a diagram like that in fig. 5.8, known to mathematicians as a
graph. The vertices or **nodes** of the graph represent the objects while the
lines or **edges** show the similarity relationships among them. Note, however, that in the graph theoretic model the nodes are not displayed in any
real relationship to space and the edges do not have either quantitative or

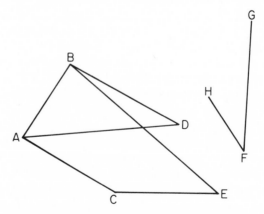

Figure 5.8 A graph (undirected); nodes represent objects and lines the presence or
absence of a given degree of similarity among them.

directional properties. A relation at any given level of similarity is either present, and represented by a line connecting two objects, or it is absent. A cluster or grouping of objects that may be of interest to the taxonomist in developing his classification is then defined as that collection of edges and nodes that exist among pairs of objects at any given value of the similarity index. For an object to belong to a cluster the minimum requirement is that it be joined to at least one other member of the cluster, although it may have connections with more than one.

The similarity coefficient itself must have properties that assure that it is monotonic—that it increases from its minimum to its maximum in a smooth, regular fashion. The following description (Estabrook 1966) provides this and other formal properties. For a set, U, having P elements (our objects to be classified) two of which are **A** and **B**, there is a similarity function, that is, a rule for assigning values, Si_{AB} which is defined for all such pairs. Values of Si_{AB} are real, finite numbers such that:

(1) The similarity of **A** with itself is equal to that of **B** with itself.
(2) The function is symmetric so that order of consideration of the two objects does not matter.
(3) The maximum similarity of an object is with itself; it may be equally but no more similar to another object.

The similarity function can be used to define what is termed in mathematics a **relation,** here designated G_c. A relation is simply some criterion that delimits a subset. In this case the name of the relation is 'linked' and we can use it to decide whether or not an object should be joined to another object or group of objects. We construct a rule that incorporates a value, c, of the similarity measure as the criterion, to wit: we will call the pair of objects **A** and **B** linked, if and only if their similarity is greater than or equal to c. G_c together with the set U is, by definition, an **undirected graph** and is designated (U, G_c). It can be diagrammed in two dimensions as the nodes **A** and **B** and the edges shown in fig. 5.8 where **A** is linked to **B,** as well as **C** to **E** and so on. Note how the properties of the model fulfill the taxonomic requirements: the criterion value of the similarity value is used in polythetic fashion to associate (link) similar taxa or specimens.

As is apparent in the illustration several G_c links may be associated forming the clusters or **linkage groups** that model the taxonomic groupings in a classification. Further, at each c level every object can be linked to one and only one group. We can define objects without c level similarity to any other object as single member groups. This means that the set U is divided into mutually exclusive and exhaustive subsets, which is also a taxonomic requirement. To define such a partitioning for each c level one must extend the G_c relationship to that of a chain of G_c or linkage relationships. A G_c chain from **A** to **F** implies that there are points **A, B,** . . . . . , **F** in U such that

$$A \; G_c \; \mathbf{B},$$
$$\mathbf{B} \; G_c \; \mathbf{C}, \; . \; . \; ,$$
$$\mathbf{E} \; G_c \; \mathbf{F}.$$

From this a second relation **A** R_c **F** can be defined when there is a G_c chain of one or more links from **A** to **F**. Further, the relation R_c can be shown to possess the following formal properties, which determine that all objects in the chain are equivalent.

(1) The reflexive property that an object is always equivalent to itself.
(2) The symmetric property that the G_c chains which are symmetric themselves can always be turned around so that **A** equivalent to **B** always implies **B** equivalent to **A**.
(3) The transitive property that if a G_c chain exists from **A** to **B** and another G_c exists from **B** to **C**, then they can be combined into a chain from **A** to **C**.

Because these properties hold, the subsets partitioning U, which are selected under R_c, are **equivalence classes.**

The graph that depicts the partitioning of U at any single level of similarity c is made up of one or more linkage groups or clusters. Taxonomically these can represent taxa; formally they are termed **subgraphs** and we say that graph (U, G_c) contains a subgraph (V, G_c) such that V contains only those points of U which are connected according to relation G_c. Such a subgraph, which is not a part of any other connected subgraph, is called a **maximal connected subgraph** (in the terminology of Estabrook [1966]) and one or more of these subgraphs are possible at any c level. Two are shown in fig. 5.8. It is these maximal connected subgraphs, then, which are the equivalence classes under relation R_c and form the partitions. The partitioning process may be repeated at decreasingly restrictive c levels (assuming that the similarity measure is monotonic) and a nested hierarchy will result. Thus, properties desirable in a taxonomic hierarchy are modeled by certain concepts from graph theory whose logical and mathematical properties are well known.

In chapter 8, we shall discuss how to draw linkage diagrams which are schematic representations of the subgraphs. These will often prove useful for discerning and conveying taxonomic information. A major feature of graph theoretic models is that it is easy to measure internal connectedness within a subgraph. Items may be connected to one or more other items. The actual number of edges among the nodes in a cluster compared with the maximum possible number provides a ratio which is a measure of internal connectedness.

In summary the graph model has properties which mimic features of the taxonomic hierarchy just as the geometric model does. The nested hierarchy and phenetic discontinuity properties are ensured by the nature of the series of equivalence relations corresponding to gradually decreasing degrees of similarity. Such relations define a partitioning, which, as we have seen, is the requirement for the taxonomic hierarchy. Discontinuities between groups are automatically defined by the criterion for membership versus nonmembership, which is always available. Further, the graph model assures absolutely that an object will not be separated from an object to which it is most similar. While this may seem to be necessary in any taxonomically useful model, in fact it is not assured in the case of some of the algorithms derived from geometric models. When distances between clusters are measured in these cases, an averaging process is often applied to the within-group object to object distances. The effect of this is to subsume individual distance measures so that subsequent group membership decisions are not dependent on all of them individually.

Besides the model for taxonomic classification, graph theory provides two other models of use in taxonomic work. Both are based on the graph theoretic definition of trees. A **tree** is a minimally connected graph in the sense that there is only one path between any two objects although it can be either direct or indirect (i.e., via intermediate objects). A **directed graph** is one in which a direction is prescribed for each edge so that the relation $B \rightarrow A$ is not the same as $A \rightarrow B$. If in a directed graph we specify a particular node, V_0, as the vertex of origin, there will be a unique path from V_0 to each other vertex. This defines a **directed tree** having its **root** at V_0. Since tree diagrams are frequently used to represent biological relationships, it will be useful to have a mathematical model for them whose properties are well known.

Two examples of the usefulness of graph theory *vis-à-vis* trees come to mind. One is the **minimum spanning tree,** which is a nondirected tree, and the other is the dendrogram, which is directed. The minimum spanning tree is simply the shortest path among all the objects. Consider the hypothetical example in fig. 5.9.

There are five objects, labeled from **A** to **E,** and the values of similarity among all possible pairs are listed. They are converted to distances, by subtracting from the maximum similarity of 1, in order to fit the graph representation of the minimum spanning tree. To connect five nodes requires four edges and the four shown as connectors (those underlined in the tables) are the shortest paths which will accomplish the task of connect-

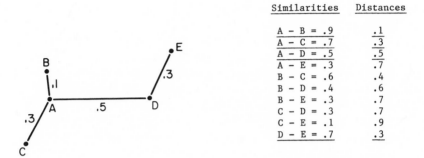

Similarities	Distances
A – B = .9	.1
A – C = .7	.3
A – D = .5	.5
A – E = .3	.7
B – C = .6	.4
B – D = .4	.6
B – E = .3	.7
C – D = .3	.7
C – E = .1	.9
D – E = .7	.3

Figure 5.9 Minimum spanning tree for 5 objects with given similarity/distance measures.

ing all of the points without forming circuits, which are precluded by definition in trees. In our case **B–C,** a shorter path than **A–D,** is not used because there is already a connection between **B** and **C** (via **A**) and the addition of **B–C** would produce a loop. The algorithm for deducing the minimum spanning tree proceeds by examining the similarities for the highest values and connecting pairs of objects that they indicate. Next, lower similarity values are examined in turn using the rule that the linkages indicated are added if and only if at least one of the objects has no previous connections. It is not a difficult algorithm to program and is contained in some computer clustering programs as the basis not only for minimum spanning trees but for single-linkage cluster analysis as well.

Another application of the tree from graph theory is seen in the **dendrogram,** often used to represent hierarchic structure (with or without phylogenetic implications). The first diagram in fig. 5.10 is an example. Note that the structure is a directed tree (admittedly upside down!) in the graph theoretic sense. It has a root at the point V_0 and a unique path from V_0 to any of the objects **A** through **D** going through either vertex V_1 or V_2, which can represent clusters of objects. The diagram is the same as the more conventional tree structure shown in the second diagram. However the first form, the dendrogram, is advantageous for representing distances or similarities quantitatively along the vertical axis.

In this example, the vertex or node, V_1, is at .9 because $Si_{AB} =$ .9, while $Si_{CD} = .85$. As shown by the connection at V_0, **A** joins **C** (as well as **D**) at the .7 level; similarly **B** joins **C** and **D** at that level. Looking at similarities among such secondarily connected pairs of objects and converting to distance measures (again by $Di = 1 - Si$) we note the following property of the distance measure in a dendrogram. If we consider distances among the triple of points **A, B, C,** then

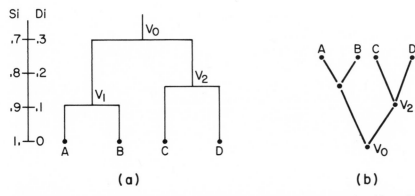

Figure 5.10 A dendrogram and a directed tree, means for representing hierarchic structure.

$$Di_{AC} = \max (Di_{AB}, Di_{BC})$$

In our case this means that the distance between **A** and **C** is .3 since that is the larger of .1 and .3. Contrast this with the metric distance measure where, in the case of three objects,

$$Di_{AC} \geq Di_{AB} + Di_{BC}$$

The maximum rule always replaces the triangle inequality property in the case of distance measures on dendrograms, and these distances are termed **ultrametrics.**

We can see now that dendrograms, as well as minimum spanning trees, are means for summarizing the information in the pairwise similarity or distance measures. Using either of them, one can trace the clustering process among objects by utilizing certain similarity or distance measures selected from those available. The linkage diagram or subgraph, on the other hand, utilizes the entire distance or similarity matrix to produce a full representation of the equivalence classes at any desired level of similarity.

Information Theory Models

We usually think of taxonomy as dealing with classification of organisms. Strictly speaking, of course, the taxonomist organizes information about organisms rather than the organisms themselves. Methods for the formal study of information have been developed in the discipline called **information theory.** Some of the results are useful to taxonomists. For one thing they have provided a useful measure of diversity within groups that corresponds to the variability within taxa (as well as to other biological phenomena like species diversity). A second important contribution to taxonomy from information theory is its provision of additional means for dealing with the simple qualitative characters that are commonly used by taxonomists but which are not amenable to many statistical treatments. We have already seen that similarity measures, such as the simple matching coefficient, can be devised for qualitative characters. Information theoretic measures add important techniques. We shall demonstrate their use in evaluating correlation among characters and diversity among objects later in this chapter. First, we must explain means for measuring information in general.

Let us consider characters whose states are defined in the multistate mode and that are not overlapping. Numerical properties are not necessarily associated with the states themselves. However, it is quite possible to associate a number with each state when that number is the number of cases described by it. This absolute frequency can be easily converted to a relative frequency by dividing it by the total number of cases. The number now associated with each state is its empirically determined probability. From a set of such probabilities, for each of the states of a character, a probability distribution can be determined for each character over the set of objects in any $n \times P$ table. As we shall see, these probabilities may be used to calculate the information in a character with respect to a particular set of data.

Ways of quantifying information contained in data are probably not intuitive to most of us. We use the very term 'information' in a specialized way: not in the sense of truth or even meaning, as in ordinary speech. We are concerned here with measuring the formal structural properties of information, not with its substantive content. In general, the amount of information carried by a particular character over a set of objects is directly expressed by measuring the difficulty of predicting what state of the character will be observed for a particular object. Thus, if it is difficult to predict, the amount learned when one actually ascertains an object's state is large; that is, relatively more information is produced. If, on the other hand, the chances are good that one can predict an object's state with respect to a character by a single guess, then the information in the character is low. (Note, again, that this does not say that the character is insignificant in terms of its meaningfulness or relevance to a study.) The relation to probability is immediately suggested by the idea of predictability, and, indeed, the information measure is a function of the probability distribution of the character states that was described above.

Information measure is the same measure as is applied to the physical quantity of entropy: the amount of disorder in a closed system. In measuring entropy we assess uncertainty. Since an uncertain system has more potentially discoverable information, large entropy measures are the same as large information measures and the two are used synonymously. Historically, the entropy measure, H, was developed first. It was later applied, first by Boltzman (1898) and then by Shannon (1948), to give us the following information measure:

$$H_a \text{ (Information in character } \mathbf{a}) = - \sum_{k=1}^{m} p_k \log_2 p_k \qquad (1)$$

where the p_k's are the probabilities of each of the 1 to m states of character **a**. (The upper case sigma, Σ, is a mathematical symbol meaning "the sum of." Here the formula means that p times $\log_2 p$ is calculated for each state in turn and the results are summed.) Although the use of logs tends to obscure the simplicity of the equation for many, the formula above is nothing more than a simple probability calculation. (Remember that summing logarithms is a way of calculating a product; here the product of the simple probabilities of the character states.)

The first thing to note about the behavior of the information measure, H, is that its value is influenced by two aspects of the distribution of character states—the number of states and the distribution of their probabilities. If all states are equally probable, the disorder, and therefore the available information, is maximal in that it is maximally difficult to predict the state of a given object. Further, that maximal H for any particular number of states can be determined, and it depends only on the number of states, k. As examples, for k of 4, H is 2, for k of 5, H equals 2.32, and for k of 8, H is 3. In fig. 5.11 we have shown the calculations.

In general, the greater the number of states, the greater the amount of information, given equally probable states. Note that for m of 1—that is, all objects in the same state of a character—H is 0. This is what one expects in that the state of an object *vis à vis* such a character is entirely without confusion.

The reason for using $\log_2$ is not intrinsic to defining information but seems convenient to us although ln (natural log or log to the base e) and $\log_{10}$, the common log, are also used. Log to the base 2 produces an information measure with units that are directly relatable to the binary bits of information stored in digital computers. Further, there is a nice interpretation of information measure computed in this way in that the maximum information in the 2, 4, 8, . . . 2^n state character is exactly equal to the number of yes or no questions that would have to be asked to ascertain the state of any object in those characters (Legendre and Rogers 1972). Since logs are readily convertible from one base to another, calculations can be done using any available log table. (For example, conversion from base 10 to base 2 is by the formula:

$$\log_2 x = \log_{10} x / \log_{10} 2)$$

(a)   Example when there are four states:

p_k = 1/4 (since all four states are equally likely)

$$H = -\sum_{k=1}^{4} 1/4 \log_2 1/4 = 4 (1/4) \log_2 1/4$$

$$= - (\log_2 1 - \log_2 4) = \log_2 4$$

By the definition of logarithms:

$H = \log_2 4$ if and only if $2^H = 4$ or $H = 2$

(b)   General formula for $\underline{k}$ states having equal probability:

$$H = \log_2 \underline{k}$$

Thus, for $\underline{k}$ = 8, $\log_2 8$ iff $2^H = 8$ or $H = 3$

and for $\underline{k}$ = 5, $\log_2 5$ iff $2^H = 5$ or $H = 2.32$

(The last case is most easily solved by converting to common logs,

i.e., $\log_{10}$, and solving for H as $\log_{10} 5 / \log_{10} 2$)

Figure 5.11 Calculation of maximal information when all states have equal probability.

These maximal measures are decreased by departures from equiprobability. The presence of a rare state is associated with increased probability, since the probability of an object's being in the common state is great. This means decreased information.

For example, for three state characters, the maximal information is 1.585. If the distribution is skewed, as illustrated in fig. 5.12, so that the probabilities of the three states are 0.5, 0.49, and 0.01, then the information is decreased to 1.0701 or to about 67% of the maximum. For the distribution 0.01, 0.98, and 0.01, it is reduced to .1615 or 10.2% of maximum, since there is a 98% probability that an object will be in state 2.

We have now described the basic information measure that applies to individual characters. These basic measures can be used to calculate a large number of derived or combined information measures that have important applications. We shall consider one that is significant in taxonomy; it provides a means for determining correlations among qualitative characters and is used in character analysis before classification.

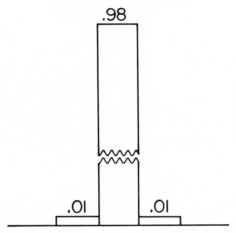

Maximum information in a three state character is $3(1/3 \log_2 1/3) = 1.5849$

For this distribution:

$$I = -\sum_{\underline{k} = 1}^{\overline{m}} p_k \log_2 p_k$$

$$= -(.01 \log_2 .01 + .98 \log_2 .98 + .01 \log_2 .01)$$

$$= .0664 + .0286 + .0664$$

$$= .1615 \quad \text{or} \quad 10.2\% \text{ of maximum information}$$

(i) For this distribution:

$$I = -(.01 \log_2 .01 + .49 \log_2 .49 + .50 \log_2 .50)$$

$$= .0664 + .5000 + .5043$$

$$= 1.0707 \text{ or } 67.7\% \text{ of maximum information}$$

Figure 5.12 Information calculations using two different distributions of character states.

Character Comparison by Information Measure

As we have seen, the general definition of correlation is a measure of the degree of association between the two members of a pair of characters. A high degree of association implies that a set of objects would be arranged or grouped in much the same way in relation to either one of the associated characters. A low degree says that knowledge about the items in relation to one character predicts little about their relationship to the other. We have already learned about the statistical measure of linear correlation, *r*, used for metric data. Now we shall consider a comparable measure that can be applied to simple qualitative data or to mixed qualitative and quantitative data when the latter is either ordered or partitioned into discrete states.

The measure is based on information theory and, like the general information measure, *H*, described above, begins with a probability distribution of the objects across the states of a given character. However, when two characters are used, other probability distributions may be defined as well. One that is suggested by the definition of correlation is the **conditional probability** distribution. Conditional probability is defined as the probability of an object's being in a state of one character given that you know it is in a particular state of another. Consider an example with characters **a** and **b** in which objects **A**, **B** and **D** are in state 2 of **b**. **B** and **D** of this group are in state 1 of **a** also. Thus, the frequency of objects in a_1 given that they are in b_2 is two out of three or $\frac{2}{3}$. This is the conditional probability and is written $P(a_1/b_2)$.

The amount of information in character **a** after observing all of its states for a given state of **b**—that is, the uncertainty in **a** that remains after removing the portion predicted by knowing a state of **b**—can be calculated in the usual way as an information measure. It is simply:

$$H_{(a/b_k)} = -\sum p(a/b_k) \log_2 p(a/b_k)$$
all *m* of **a**

where $p(a/b_k)$ is the probability of one state of **a** given the kth state of **b**. Note that the function of the probabilities of each state of **a** are calculated and the results summed. The same procedure is then applied in turn to other states of character **b** and the weighted sum computed as

$$H_{(a/b)} = \sum_{k=1}^{m} H_{(a/b_k)} p(b_k)$$

This information, $H_{(a/b)}$, can be interpreted as information held exclusively by **a** with respect to **b**, since it is the information remaining in **a** after what can be predicted by observing **b** has been accounted for.

The example in fig. 5.13(a) illustrates these rules as applied to part of the data used in chapter 2 from Legendre's study of *Salmo* (1972). We employ values for ten of the specimens on the three characters:

(a) Pyloric Caeca Number: State 1 = 17–26, State 2 = 27–35, State 3 = 36–44, State 4 = 45–60.

(b) Color of Cutthroat Marks: State 1 = red, State 2 = yellow, State 3 = blank (no marks).

(c) Size and Location of Spotting: State 1 = large, posterior, State 2 = medium, posterior and more anterior, State 3 = fine and profuse, generalized.

To determine the correlative relationship between Pyloric Caeca Number (variable **a**) and Color of Cutthroat Marks (variable **b**) we would find the information for each of the four states of **a** corresponding to each of the three *(k)* states of **b**. These probabilities are recorded in fig. 5.13(b) in the column marked "Conditional Probability of **a/b**," which shows, for example, that in the three cases with red cutthroat marks two have caeca numbers in State 1 while the other is in State 2. The probabilities for each of the three states of cutthroat marks are shown in the lowest row of the table. A similar table is shown in fig. 5.13(b) for the conditional probability of **b/a**. Below the table are the computations for the information measures of the conditional probabilities. In each case the calculations were done in two steps. First, information measures were computed, conditioning on each state of the variable separately (as, for example, information in Pyloric Caeca Number *given* red Cutthroat Color). Second, these measures were summed after being weighted by the probabilities of each of the states.

$H_{(b/a)}$ is the information held exclusively in character **b**, the Color of Cutthroat Marks, while the first calculation, $H_{(a/b)}$, is that held exclusively

Specimen No.	101	201	210	301	403	406	507	710	806	801
a Pyloric Caeca Number	4	2	2	1	3	3	2	4	2	1
b Color of Cutthroat Marks	3	2	2	1	3	3	2	2	1	1
c Size and Location of Spotting	3	3	3	2	3	3	2	2	1	1

Figure 5.13 Calculations for character comparisons using information measures in *Salmo* study: (above), data table, $n \times P$ with 3 characters and 10 objects; (following 2 pages), information measures based on conditional probabilities; (following third page), information measure based on joint probabilities.

Conditional Probability of a/b

Variable b

States	1 red	2 yellow	3 blank
1 (17 – 26)	2/3	0	0
2 (27 – 35)	1/3	3/4	0
3 (36 – 44)	0	0	2/3
4 (45 – 60)	0	1/4	1/3

a (left label)

$p(b_k)$ 3/10   4/10   3/10

Step 1

$H_{a/b} = - (2/3 \log_2 2/3 + 1/3 \log_2 1/3 + 0 + 0)$
$= .92$

$H_{a/b} = - (0 + 3/4 \log_2 3/4 + 0 + 1/4 \log_2 1/4)$
$= .81$

$H_{a/b} = - (0 + 0 + 2/3 \log_2 2/3 + 1/3 \log_2 1/3)$
$= .92$

Step 2

Since $p(b_1) = .3$, $p(b_2) = .4$, $p(b_3) = .3$

$H_{a/b} = .3(.92) + .4(.81) + .3(.92) = .876$

Conditional Probability of b/a

Variable a

States	1 (17–26)	2 (27–35)	3 (36–44)	4 (45–60)
1 red	2/2	1/4	0	0
2 yellow	0	3/4	0	1/2
3 blank	0	0	2/2	1/2
$p(b_k)$	2/10	4/10	2/10	2/10

b

Condensing the steps into one to find $H_{b/a}$

$$H_{b/a} = 2/10 \; (0) + 4/10 \; (1/4 \; \log_2 1/4 + 3/4 \; \log_2 3/4)$$
$$+ \; 2/10 \; (0) + 2/10 \; (1/2 \; \log_2 1/2 + 1/2 \; \log_2 1/2)$$
$$= \; .524$$

(Note: No information in Pyloric Caeca States 1 and 3,
 each is invariant with respect to Cutthroat Marks
 and, therefore, predictable with certainty.)

<table>
<tr><th></th><th></th><th colspan="4">a
Pyloric Caeca Number</th></tr>
<tr><th></th><th>States</th><th>1</th><th>2</th><th>3</th><th>4</th></tr>
<tr><td rowspan="3">b
Cutthroat Marks</td><td>1</td><td>2/10</td><td>1/10</td><td>0</td><td>0</td></tr>
<tr><td>2</td><td>0</td><td>3/10</td><td>0</td><td>1/10</td></tr>
<tr><td>3</td><td>0</td><td>0</td><td>2/10</td><td>1/10</td></tr>
</table>

$$
\begin{aligned}
H_{(a \cdot b)} &= .2 \log_2 .2 + .1 \log_2 .1 + .3 \log_2 .3 \\
&\quad + .1 \log_2 .1 + .2 \log_2 .1 + .1 \log_2 .1 \\
&= 2.45
\end{aligned}
$$

in the Pyloric Caeca Number character. We can also determine the information possessed by both. The latter is the information measure based on the **joint probabilities;** that is, the probability of all possible pairwise combinations of character states. This distribution for our example is seen in fig. 5.13(c). Note that in the case of joint probabilities each is calculated over the total number of objects unlike conditional probabilities, which are based on a reduced set of objects corresponding to the number in the state conditioned on. Again, the information function is applied to joint probabilities in the usual way using the following formula:

$$
H_{(a \cdot b)} = - \sum_{k=1}^{m_a} \sum_{h=1}^{m_b} p(a_k \cdot b_h) \log_2 p(a_k \cdot b_h)
$$

Calculations are shown in fig. 5.13(c) where $p(a_k \cdot b_h)$ is the probability of an object in state k of **a** and state h of **b** simultaneously in a particular set of objects. The double summation sign ($\Sigma \ \Sigma$) indicates that one sums over all the states of **b** for each state of **a** and then moves to the next state of **a** and sums over all of **b** again, and so on. Since there is summation over all states of both characters, the measure represents the total information in both, that is, that held exclusively in each plus that shared. We note that because there is shared information, the information in both characters ($H_{(a \cdot b)}$) is not simply the sum of the information in each separately. Various measures and coefficients concerning pairs of characters can be derived from the four basic information measures: H_a, H_b, $H_{(a/b)}$ and $H_{(b/a)}$, and $H_{(a \cdot b)}$. They will be described and their usefulness discussed in chapter 6.

The reader may have noticed that all of our information measures have been based on a distribution of objects per character state. That is to say, we are doing a row analysis of our $n \times P$ table; this is commonly referred to as **R analysis**. The same kind of analysis could be applied to the columns of the table by counting the number of like character states per object. This is called **Q analysis** and is commonly performed in ecological work. In typical ecological cases the sites are the items and a series of species the characters (often called attributes in the ecological literature). Presence or absence of each species or a number or a descriptor state like rare or abundant, which summarizes its frequency in a site, is recorded in the body of the table. In this case one could measure information according to sites per species (R analysis) or as species per site, which is Q analysis. In the former case one is looking at rare versus common species over sites. In the latter the question is whether a site has many species or few and it makes sense to interpret the relationship among sites as a similarity measure and to classify sites according to like or different species.

Interpretation of such a Q analysis in the case of most taxonomic data tables would be more difficult than in the above ecological situations. There are a number of problems: for one, the coding of the states is arbitrary for simple (unordered) multistate characters. Further, the same set of character states is not usually scored for each character, as is done for species in ecological studies where, for example, a set of states like "rare," "common," "abundant" is coded as 1, 2, 3 for many species per site. By contrast in taxonomic data, the number of 3s for object **A** is likely to be meaningless and simple recoding can change the probability distribution.

Information Measures of Diversity

A different approach, which is particularly useful for taxonomic classification, applies information analysis to sets of objects rather than to pairs of characters. Here we measure the degree of homogeneity within the group directly. Let us say we have a group of J objects and look at their distribution over a character. If the objects are evenly distributed over the states of the character, the information measure for that character will be maximal, while if there are rare states and common states, the information in the character will be reduced as we have seen.

Information measures of this type have the fortunate property of being additive over the characters, under the proper conditions. This makes it possible to compute the overall information in any $n \times P$ table or any subset of objects in such a table. These overall information measures are referred to as diversity measures and will be designated by I. The use of the term diversity is analogous to its use in ecology with reference to species diversity, for example. Diversity measures are subject to the important constraint (Pielou 1969) that the characters must be independent. Thus, for n independent characters the diversity is:

$$I = \sum_{i=a}^{n} H_i \quad \text{(where } H_i = - \sum_{k=1}^{m} p_k \log_2 p_k)$$

An example of the calculation for the whole group of ten objects over three characters using our *Salmo* data is seen in fig. 5.14(a). The result is the average diversity per individual, since proportions are used in the formula. Total diversity for the group of ten objects would equal this times ten.

    Total diversity may be calculated directly from a different, but

Total information in Pyloric Caeca Number is found by applying formula (1) to the probabilities of its states as shown in Fig. 5.13 as:

$$H_a = - (.2 \log_2 .2 + .4 \log_2 .4 + .2 \log_2 .2 + .2 \log_2 .2)$$
$$= 1.921$$

Similarly, information in Color of Cutthroat Marks, H_b , is given by:

$$H_b = - (.3 \log_2 .3 + .4 \log_2 .4 + .3 \log_2 .3)$$
$$= 1.571$$

For Size and Location of Spotting, using probabilities of its states from data in Fig. 5.13:

$$H_c = - (.2 \log_2 .2 + .3 \log_2 .3 + .5 \log_2 .5)$$
$$= 1.485$$

Summing these gives the information measure of diversity of the group as:

$$I = \sum_{i=1}^{n} H_i = 1.921 + 1.571 + 1.485 = 4.98$$

$$I = 3 (10 \log_2 10) - (2 \log_2 2 + 4 \log_2 4 + 2 \log_2 2 + 2 \log_2 2)$$
$$+ 3 \log_2 3 + 4 \log_2 4 + 3 \log_2 3$$
$$+ 2 \log_2 2 + 3 \log_2 3 + 5 \log_2 5)$$
$$I = 99.658 - (2 + 8 + 2 + 2 + 4.75 + 8 + 4.75 + 2 + 4.75 + 11.6)$$
$$I = 49.8$$

Figure 5.14 Information measures of group diversity: (upper) calculation of average diversity per individual; (lower) direct calculation of total diversity.

commonly used, form of the information measure. The information in each character is given by:

$$H_i = N \log_2 N - \sum_{k=1}^{m} n_k \log_2 n_k$$

Summing over t characters to give I is written:

$$I = t N \log_2 N - \sum_{i=1}^{t} \sum_{k=1}^{m} n_k \log_2 n_k$$

In both of these formulations n_k refers to the number of objects in character state k while N is the total number of objects. That the total diversity is the same as that from the first formula is demonstrated by the calculation in fig. 5.14(b).

Once we know how to compute diversity for a set of items over any number of characters, we can apply the measure to various subsets of the objects. Let us say we have three subsets and we wish to combine the two of them that are the most similar. In this case we shall determine degree of similarity by measuring the change in diversity that occurs upon fusion of two groups. Fusion of two nonidentical groups will necessarily result in a new group with more diversity. However, the two groups that are the most similar will have the least change in diversity or the minimal **information gain,** a quantity that we shall denote as ΔI. The procedure is to calculate ΔI values for all possible pairings of groups (or individuals). In our example of combining two of three groups, three ΔI's must be computed. We then choose the fusion with the minimum increase in information.

There are several ways of computing ΔI values. Different ones are suitable in different situations and for different data types (Orlocci 1969a and 1969b, Clifford and Stephenson 1975). We shall present only one. It is a simple comparison of the information in the fused groups with the sum of the information in the separate groups. Thus:

$$\Delta I = I_{X+Y} - (I_X + I_Y)$$

where X and Y are either groups or individual objects. (The latter will always contain zero information since there can be no diversity of states.)

Using our example again from fig. 5.13 and the second form of the diversity equation, we could compute the information gain resulting from fusing a group, X, containing objects 301, 507, 801, and 806 with group Y which contains 210, 403, and 406. Calculations are shown in fig. 5.15.

We see that groups of objects which share rare states, and therefore have a lower information content, will be rated as more homogeneous

$$I_X = 3 \ (4 \ \log_2 4) \ - \ [(2 \ \log_2 2 + 2 \ \log_2 2 + 0 + 0)$$
$$+ \ (3 \ \log_2 3 + \log_2 1 + 0)$$
$$+ \ (2 \ \log_2 2 + 2 \ \log_2 2 + 0)]$$
$$= 11.25$$

$$I_Y = 3 \ (3 \ \log_2 3) \ - \ [(0 + \log_2 1 + 2 \ \log_2 2)$$
$$+ \ (0 + \log_2 1 + 2 \ \log_2 2)$$
$$+ \ (0 + 0 + 3 \ \log_2 3)]$$
$$= 5.51$$

Computing the information in the fused group is done as follows:

$$I_{X+Y} = 3 \ (7 \ \log_2 7) \ - \ [(2 \ \log_2 2 + 3 \ \log_2 3 + 2 \ \log_2 2)$$
$$+ \ (3 \ \log_2 3 + 2 \ \log_2 2 + 2 \ \log_2 2)$$
$$+ \ (2 \ \log_2 2 + 2 \ \log_2 2 + 3 \ \log_2 3)]$$
$$= 32.68$$

Thus, Δ I $= 32.68 - (11.25 + 5.51) = 15.9$

Figure 5.15 Calculation of change in information.

by the information gain approach. Thus, they will be joined more readily. Whether or not this is a favorable property with respect to taxonomic analysis is arguable. In one sense it parallels common practice in that unique, usually rare, states are often sought out as key characters for identifying taxa. On the other hand, as Sneath and Sokal (1973) point out, the information gain procedure is equivalent to weighting characters that have rare states. In their view, which generally champions use of unweighted characters, this is a major disadvantage of information analysis in taxonomic work.

Algorithms for Producing Taxonomic Hierarchies

All of the models described can be used to describe hierarchical organization of taxonomic objects. The geometric model is perhaps the most intuitive and therefore the one most frequently assumed by people. An often quoted advantage of set/graph theory and information theory models is that they do not make any assumptions about linearity of the relationships in the data. Further, they are not dependent on pre-determined distributions, such as the normal curve, which underlie some uses of the geometric model.

In any case, no matter what choice is made among the models we will need an algorithm—that is, a means for actually deriving the clusters, identifying the members of each. Often this will require assistance from a computer. The methods we use may or may not be closely related to the model used.

Algorithms are the basis for writing computer programs. We shall not consider here the programming aspects of the procedural problem, although in practice they can be critical. More than one algorithm may be devised for a particular purpose. Choice among them is frequently made on the basis of computability. Obviously, computer time and cost will depend on the amount of data. Very large numbers of objects may make one algorithm preferable to another. Thus, choice of algorithm is a practical problem of finding a strategy which suits the model, the problem, and the data and is efficient to compute.

Except in the case of ordination, clusters in a hyperspace cannot be determined by visual inspection. Instead we must use the multivariate distance or similarity measures we can compute to reveal the hierarchic structure in the data. There are basically three strategies which can be invoked, either alone or in combination, to determine clusters. One is the **agglomerative** procedure, in which groups are formed from the bottom up, starting with small groups of the most similar objects and gradually adding to them less similar objects or groups. Thus, group membership grows by addition of new objects and fusion of preexisting groups until all objects are joined in a single large group. A second approach to finding hierarchic structure in a set of objects is **divisive.** Here we proceed from the top down—the whole set is repeatedly subdivided. A third important type of algorithm for discovering the hierarchic structure in a data table is **optimization.** Such an algorithm proceeds by some degree of trial and error; that is, several proposed subsets are compared according to a pre-set criterion. The subset that optimizes—that is, either maximizes or minimizes—the criterion is chosen. For instance, one might seek to minimize information gain or maximize the between-group sums of squares as we shall see below.

The characters that are available, and whether they will be used as multiple variables simultaneously or sequentially, has an important influence on the choice of strategies. Agglomerative techniques are best adapted for using data to form polythetic groups or classifications—those in which there are large numbers of shared character states and no single one is required for membership in the group. Monothetic groups are generally associated with divisive methods, although there are a few algorithms for polythetic divisive classifications. In addition to these general relations to character use, algorithm choice is mediated both by the data type and by the model preferred. Let us look at a number of common model–algorithm combinations.

Agglomerative Polythetic Algorithms

We have seen that pairwise measures among objects are associated with both the geometric and graph models. They are distances in the geometric case and similarity values in the graph model. Distance measures range upward from zero for identical objects; similarity measures indicate identical objects by a value of one and apply zero to the maximally dissimilar pair. The two measures can generally be converted from one to the other. More importantly, both are amenable to the same algorithm for cluster determination— the agglomerative polythetic. It is quite clear that both distance and similarity measures employ characters in a polythetic fashion—that is, all characters are applied simultaneously to derive the degree of sharing of character states, which is the similarity value. It is also easy to see that the agglomerative method is suggested by the graph model. Groups are defined as equivalence classes in that they contain members at least as similar as a criterion level called c. By beginning with the c values for the most similar pairs of objects in a data set and gradually relaxing the degree of similarity required for group membership, a series of ever larger groups is identified by means of an agglomerative procedure. If similarity values are used, the c level begins at one and decreases by steps toward zero. If distances are used, the c level will increase from zero. The direction of c change is not critical but it must go continuously in one direction—that is, monotonically. This means that once objects have been joined at level c, all later joinings will occur at lower levels of similarity (or greater distance).

    The graph model proposes a simple rule, known as the **single-linkage** rule, for adding a new member to a group: the new member proposed must have a certain, stated criterion level of similarity to at least one member of the group. There are numerous other rules and the cluster memberships determined vary widely depending on the rule used. In general the rules for agglomerative clustering based on pairwise measures fall into two types. The first are extensions of the single-linkage, or nearest neighbor rule, as it is sometimes called. An example is the furthest neighbor rule in which an object must be similar at the criterion level, c, to all other objects in a group in order to join. Such rules continue to employ only the pairwise similarities in the original $P \times P$ matrix and require no further calculations.

Other rules are suggested by the geometric model and require new calculations of similarity at each step in determination of cluster membership. For example, average distance between objects in a group and a proposed member may be calculated and the object joined according to the shortest mean distance. Or centroids of existing groups can be found and groups fused when distances between their centroids are less than c. Many of these rules have been used in taxonomic work. There are theoretical advantages to the single-linkage rule, since its mathematical properties are known (Jardine and Sibson 1971).

Agglomerative Polythetic-Optimization Algorithms

There are agglomerative cluster analysis algorithms that function on group properties rather than pairwise comparisons of objects. As we have seen, the diversity measures from information theory is such a group measure. The information gain ΔI described above, can be regarded as a function indicating distance between groups in the sense that combining two groups that are nearly similar gives a lower ΔI that corresponds to a small distance. Since groups, unlike objects, do not preexist the classification process, the inter-group distances required by this strategy must be calculated at each step. Further, optimization is necessary in that all possible cluster formations must be surveyed in order to determine the one which produces the minimum information gain. However, the combination of information gain distances between groups and minimization is a useful agglomerative polythetic algorithm for hierarchical clustering, particularly when there are relatively few objects (since computing times increase in some exponential proportion to the number of objects) and characters are simple multistate including symmetric binary (Clifford and Stephenson 1975). A similar strategy makes use of minimal increase in sums of squares upon fusion of groups. It makes use of the constant mathematical relationship (Burr 1970) between the average of the squared Euclidean distance among all possible pairs of objects over a given character and the sample variance for that character.

Monothetic Divisive Algorithms

Divisive classification strategies operate on the full set of objects, first dividing it into two or more subsets according to some criterion. The next step is to divide the resulting subsets, and so on. The last step in the logic would see all individual objects separated, but it is not usually desirable to proceed to this ultimate stage. In fact, one of the major theoretical advantages of a divisive algorithm is that it identifies the major groupings without the noise that may be introduced by the less easily distinguished small groups.

While divisive grouping strategies are multivariate, the variables are not used all at once, as they are in computing similarities where their combined effect is expressed by a single number. Instead the multiple variables are used sequentially. This leads, naturally, to a monothetic classification. Since groups are separated in relation to the states of one descriptor at a time, all members of a group must be described by the same series of descriptor states.

The major problem for divisive strategies is, then, to determine the sequence of application for the descriptors or variables. There is the logical possibility of choosing the sequence on the basis of *a priori* biological knowledge of the group. For example, for the phylogenetic classification it would make good sense to use characters in the order of their primitiveness if this were known, which is not likely. Computer-assisted optimization procedures are more commonly employed and *a posteriori* decisions made with reference to the data set itself. The general idea is to form subsets of the objects on the basis of each character in turn and to pick that one which optimizes some property of the resultant groups.

We shall describe two examples briefly. Association analysis (Williams and Lambert 1959) is one of the best known. It is applicable to binary data and was once widely used for ecological problems. The chi-squares of each 2-state character compared with every other character are summed and the largest one chosen for dividing the objects into two subsets on the basis of the two character states. These two subsets are now further subdivided on the same basis as before using the remaining characters. Later it became known (Lance and Williams 1968) that this process also defines an information split in that measures of information are related to the estimate of chi-squares. This suggested the strategy based on the information measure ΔI, where the total information content of the group on one character is first calculated and then the group is divided into two subsets and the information content of each calculated. $\Delta I = I_{X+Y} - (I_X + I_Y)$ is found as before, but this time maximal fall in information is required to indicate the character to be used as the basis for division.

A major problem of monothetic classification routines is the tendency to generate obvious misclassifications. If an object is like all others in a group except for one character state, it may be from then on barred from that group if that is the character used as the criterion. It is easy to see that early errors will be perpetuated. For this reason monothetic programs need to be provided with the means for reallocation of objects. In general, divisive algorithms are not widely used in taxonomic work for producing classification. In addition to the misclassification problem they tend to be too rigid to deal well with the lower taxa, which is where many of the problems in taxonomy lie. However it should be noted that the construction of keys is an example of the use of monothetic, divisive algorithms.

There have been a few polythetic divisive algorithms devised. Theoretically, they would be ideal since divisive programs deal better with large groups of objects than do agglomerative ones. When metric data are available one can compute the principal components and divide objects into pairs of subsets on the basis of each component in turn, so as to maximize the between-groups sums of squares.

Suggested Readings

General

Dunn, G. and B. S. Everitt 1982. *An Introduction to Mathematical Taxonomy*. Cambridge: Cambridge University Press.

Jardine, N. and Sibson, R. 1971. *Mathematical Taxonomy*. London: Wiley.

Sneath, P.H.A. and R. R. Sokal 1973. *Numerical Taxonomy*. San Francisco: W. H. Freeman.

Information and Cluster Analysis

Clifford, H. T. and W. Stephenson 1975. *An Introduction to Numerical Classification*. New York: Academic Press.

Everitt, B. S. 1980. *Cluster Analysis*, 2nd Edition. London: Heinemann.

Legendre, P. and Rogers, D. J. 1972. Characters and clustering in taxonomy: A synthesis of two taximetric procedures. Taxon 25(5/6): 567–606.

Multivariate Analysis

Marriott, F.H.C. 1974. *The Interpretation of Multiple Observations*. London and New York: Academic Press.

Pimentel, R. A. 1979. *Morphometrics: The Multivariate Analysis of Biological Data*. Dubuque: Kendall/Hunt.

Part III

Computer-Assisted Taxonomic Analysis

Chapter 6

Character Analysis

This chapter should be read in connection with chapters 7 and 8: here we discuss the analysis of characters for use in phenetic pattern analysis and classification; in chapter 7 we discuss the methods of phenetic pattern analysis and classification themselves; and in chapter 8 we look at methods of depicting the results. Other aspects of characters relating to identification and cladistics are covered in chapters 9 and 10. The reader will have met the taxonomist's concept of characters and conventional methods for defining and using them in chapter 3.

Classificatory Value

The classificatory value of any taxonomic character lies primarily in its distribution among the items and the correlation of this distribution with those of other characters. In a frequently quoted passage from chapter 13 of the *Origin of Species,* Darwin (1859) writes that "The importance, for classification, of trifling characters, mainly depends on their being correlated with several other characters of more or less importance." This principle has four consequences which have not always been appreciated:

1. The value of a particular character at a particular level of the taxonomic hierarchy can be determined only *after* it has been observed and recorded. Hence the advocacy of weighting or evaluation *a posteriori* ("in arrears") rather than *a priori* ("in advance") (Davis and Heywood 1963; Legendre 1975).
2. Characters of high value in taxonomy may be based on features of large or small biological consequence. Here we may compare some of the impor-

tant characters that distinguish the Orobanchaceae from the Scrophulari-
aceae. The fact that the former are parasites (lacking chlorophyll, photo-
synthesis, and developed leaves) is used alongside seemingly inconsequential
differences between them in ovary structure.

3. Characters of high value may be characters whose functional significance
is well known or poorly known. Thus in comparing the Monocotyledons
with the Dicotyledons, we may make confident statements about the roles
of different vascular arrangements and primary roots, but find ourselves
unable to comment on other characters of taxonomic merit, such as the tendency of Monocotyledons to bear flower parts in multiples of three.

4. Any particular character may prove valuable at one taxonomic rank in one
group, at another rank in a second group, or even of no value in a third.
The distinction between flowers with superior and inferior ovaries is an example: it provides the diagnostic distinction between Liliaceae and Amar-
yllidaceae, varies between subfamilies in the Rosaceae, and is so variable
as to be of little use in the genus *Saxifraga*.

Types of Characters

Taxonomists use the term character very loosely to cover the comparison of
homologous conditions at any stage in a taxonomic study. We distinguish
here a variety of stages through which characters evolve: from the crudest
comparison or recording of observations, through several intermediate stages,
to the publicized characters by which taxa are circumscribed or identified in
monographs, Floras, and Faunas published at the end of a study.

Features and Raw Observations

We should note that characters or character formulations are human constructs. The plants themselves bear **features,** which are the product of interactions between the plant genotype and the environment (Allkin 1979). Before formulating characters taxonomists must make **raw observations** of
features and then connect these together, deciding which are homologous and
what character states or values are to be compared (Bisby 1970; Bisby and
Nicholls 1977).

Crude Characters

Crude characters are the preliminary definitions of characters sufficient solely
for recording purposes. In general they involve detailed continuous measure-

ments or the recording of many detailed character states in a multistate character. Subsequently the detail may be reduced once a satisfactory working character has been defined, but to start with one has to err on the detailed side and then refine the character.

Working Characters

If a character definition is to contain states or measures suitable for absolutely all conditions observed, one would first have to observe all the material and then define the characters. In practice what happens has two stages. First, taxonomists look at a small sample of the material and score the crude characters which come to mind. As far as possible they arrange for the sample to contain a wide range from the known variation in the material. At this stage there is also a clear bias toward characters and states which come to mind because they were used by taxonomists in earlier treatments. For the second stage, the taxonomist does a preliminary selection and redefinition of crude characters to produce **working characters** for use throughout the remainder of the observation work. A working character must be a putative version of a final character—with a level of detail and definition of states or units that it is hoped will be of use once the data are complete. It must be borne in mind that new features may be discovered among the items. This will cause the introduction of new states, new characters or, most difficult of all, changes to the definition of states. This latter produces the problem of having to go back to rescore materials already scored once. So the working characters may remain unchanged throughout the scoring process or they may be supplemented or changed. It is, however, very important that the taxonomist allow them to evolve, particularly because of the initial bias toward distinctions made by earlier taxonomists.

Final Characters

When scoring is complete the taxonomist has the final version of the scored characters. Even now they may suffer further selection and further changes for the particular use that will be made of them. We are concerned here with their use as classificatory characters, but other versions, what Allkin and White (1982) call **sister characters,** will be used as descriptive characters, key characters, cladistic characters, or retrieval characters. It is at this final stage that computer aids are of importance in analyzing, selecting, or defining characters. In theory the techniques might be of use in processing characters before a conventional taxonomist does the pattern analysis, but in practice they tend to be used prior to further computer analysis of the type described in chapter 7. In general the computer analyses require more precise character formulations than do conventional techniques, which is what forces those using computer techniques to study their characters more precisely.

Structural Properties of Classificatory Characters

There are many different views on the extent to which taxonomists carrying out a computer-aided classification should investigate both the biological and structural properties of their characters. We feel strongly that they should pursue both as far as is practicable. In the first place it is essential to learn as much as one can about the biology so as to achieve high standards with homology and character state definition. Secondly, characters of given distributions have known operational advantages and disadvantages for particular purposes. Some numerical taxonomists argue against acquiring this information, supposing that intervening in the character definition, knowing that this may have an effect on the results, is somehow cheating. In our opinion the notion that characters ought to be applied blindly, in ignorance of each character's effect on the collective pattern, is both naïve and, among dedicated taxonomists, unrealistic.

Distribution of Character State Frequencies

The first information about a working character needed after scoring it is the frequency of occurrence of each state (for qualitative characters) or the frequency distribution (for quantitative characters). The taxonomist will want to know whether all states were actually used and in how many items the character was not scored because of missing or inapplicable observations. Among those states that were used one will want to know the distribution, if only to categorize the character crudely on the scale between rare-state and balanced characters. For quantitative data one will want to see whether the distribution is multimodal, and if so whether there are gaps in the distribution.

Amount of Information in a Qualitative Character

The amount of information (H, see chapter 5) in a qualitative character can be calculated from the scored data for that character for a given set of items.

However, characters with a high information content are not necessarily of great value in producing classifications. Indeed, one can exclude characters having extreme information contents. A character for which $H = 0$ must be invariant (i.e., all taxa have the same state) for all the items and hence be of no use in grouping them. Alternatively a character with as many different states as items has a high information content, but is again of no use in grouping, although it may be ideal for identification. So a minimum requirement is that a character partition the items into two or more states, with at least one state held by more than one item.

Even within this definition we can distinguish some commonly occurring extremes: **balanced characters,** in which a few states occur with about equal frequency; and **rare-state characters,** in which one or more states occur rarely, perhaps only once. All other considerations being equal (and they frequently are not; we shall discuss below the importance of taxonomic correlations) balanced characters produce balanced classifications, whether analyzed traditionally or by numerical means. For instance characters no. 8 and no. 5 in table 6.1, showing morphological characters (Keel Shape: "curved" (158) or "angled" (115), and Crest on Keel: present (119) or absent (153)) would, whether or not they were correlated, contribute to a few large groupings of *Crotalaria* species. In contrast, the use of rare-state characters often contributes to the segregation of small groups of oddities. Thus characters no. 3, 35, 43 and 28 in table 6.1 did in fact contribute to the recognition of small sections in *Crotalaria*. For instance the single species with a split stigma is placed in section Schizostigma.

It is not suggested that taxonomists should actively seek one particular kind of character, rather that they should be aware of the conse-

Table 6.1 The Characters Analysed, Listed in Descending Order of Information Contribution

Character no. as used in the text and computations	Information contribution	Character description	Frequencies of occurrence of character states*	
4	6.14	Anther shape	231:42	p
3	5.27	Stigma shape	272:1	p
57	4.84	Filament length	219:54	Bi
40	4.65	Position of the inflorescence	214:16:43	p
25	4.55	Mature calyx deflection	229:38	
15	4.32	Standard appendage attachment	149:118:5:1	PP
2	4.19	Style shape	3:157:112:1	p
10	4.08	Beak twisting	155:24:93	PP
8	4.04	Presence of crest on keel	119:153	PP
5	4.00	Keel shape	158:115	p
29	3.82	Markings on petals	221:52	p

Table 6.1 (*continued*)

Character no. as used in the text and computations	Information contribution	Character description	Frequencies of occurrence of character states*	
38	3.41	Bract persistence	28:197:47	
48	3.41	Tree or not	15:258	
1	3.37	Style pubescence	7:138:114:14	p
24	3.33	Calyx lobing	150:99:24	PP
35	3.20	Legume inflation	264:2	
19	2.75	Seed thickness	236:6:1	
17	2.55	Seed number	44:49:174	p
22	2.45	Hypanthium development	235:38	p
34	2.42	Legume shape	56:39:98:72	p
43	2.38	Presence of spines	266:7	Ba
6	2.32	Keel indumentum	249:15:8	p
21	2.31	Aril development	205:37	p
31	2.27	Stipe attachment	227:34:5	p
36	2.24	Bracteole position and form	30:113:17:96:17	p
28	2.24	Calyx lobe number	264:9	
37	2.05	Bud inflection	122:98:53	p
26	1.95	Calyx lobe shape	152:103:13:5	
32	1.95	Stipe/calyx tube length	45:55:30:134	
27	1.77	Calyx lobe/tube length	194:79	
14	1.69	Adaxial indumentum of standard	188:68:17	
46	1.65	Development of petioles	31:242	
47	1.65	Development of stipules	113:141:19	
23	1.55	Calyx/keel length	156:48:69	
13	1.51	Shape of standard	53:108:38:21:53	
7	1.48	Keel upper margin indumentum	75:184:14	
9	1.43	Keel length	125:120:28	Ba
52	1.40	Prominence of lateral ridges on keel	37:234	Bi
30	1.39	Flower colour	21:251	p
44	1.33	Leaf-form	24:22:227	Ba
33	1.33	External indumentum of legume	68:172:23	
51	1.27	Reflexing of calyx lobes	253:20	Bi
12	1.26	Wing/keel length	202:20	
20	1.16	Seed surface	183:60	
45	1.15	Development of leaf fascicles	235:38	
41	1.14	Raceme form	200:53:20	
18	1.11	Seed shape	20:210:16	
42	1.07	Flower number per inflorescence	81:192	
50	1.04	Seed colour	5:127:70:7:6	Bi
16	0.93	Internal indumentum of legume	241:23	Ba
11	0.80	Wing shape	72:200	
39	0.75	Bract/pedicel length	64:209	

Source: Bisby 1970, table 1.

*these frequencies total 273 except where there is missing data.

quences of using the different types. Decisions on whether to use rare-state characters and therefore possibly recognize segregate groups will depend largely on their level of correlation. In the legume tribe Genisteae the species *Spartium junceum* (Spanish broom) is unique in many features: its keel is a distinctive shape, its anthers are spinose, its pod partitioned, and its calyx strap-shaped. All of these are rare-state characters correlated in such a way as to emphasize its distinctness and hence encourage its delimitation as a separate genus *Spartium*.

Distributions of Quantitative Characters

When a taxonomist starts recording a quantitative character for a set of taxa the distribution curve to expect is not known. After scoring the character one can draw the frequency distribution curve and note whether it is unimodal or multimodal, whether it is symmetrical or skewed, and whether it shows discontinuities. Particularly with taxa that are above the species level, a variety of curves—and not always normal curves—may be encountered.

Characters with multimodal or discontinuous frequency curves are more useful in making classifications than characters with normal curves. The reason is this. Where a character with a discontinuous frequency curve is taxonomically correlated with a second which also has a discontinuous curve the result is a most clearcut discontinuity. For example fig. 6.1 shows the histogram of the heterophylly index (ratio of upper leaf length divided by breadth to lower leaf length divided by breadth) for vetches in the *Vicia sativa* aggregate from Hollings and Stace (1978). Hollings and Stace found

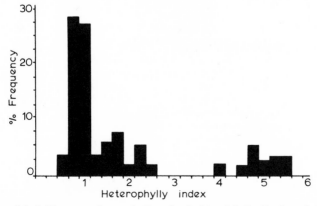

Figure 6.1 A histogram demonstrating clearly bimodal distribution of character states in a vetch. *(From Hollings and Stace 1978. Fig. 2).*

that this rather complicated measure, which shows the change in leaf shape on one plant, was taxonomically correlated with another character, petal color, with a similarly discontinuous distribution. As a consequence the combination of these two characters was selected for the major distinction between two pairs of very similar species. *Vicia lathyroides* and *Vicia angustifolia* (both with strongly varied leaves and petals of a uniform color) were thus separated from *Vicia segetalis* and *Vicia sativa* (both more uniform in leaf shape and with two-colored petals). In contrast the unimodal distribution of pod-lengths shown for the same plants in fig. 6.2 was of much less use: it did not imply any discontinuity and indeed the species recognized on other characters had overlapping ranges for this character.

Inspection of the mixed-data similarity measures, described for use with mixed quantitative and qualitative data in the next chapter, will reveal that differences between values near the mean receive relatively lower weight than other differences and so contribute little to the measure of overall similarity computed. But for a character with an approximately normal distribution, such comparisons will be the most common. In contrast, differences between extreme values occur more frequently with discontinuous,

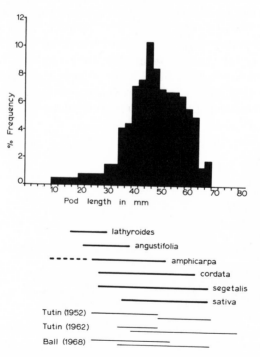

Figure 6.2 A histogram showing the unimodal distribution of pod length in a vetch species. *(From Hollings and Stace 1978. Fig. 5).*

bimodal, or multimodal characters and so they contribute relatively more to the result. Greatest relative weight is contributed to these coefficients by binary characters where the similarity is "all or nothing" with no partial values.

This "unequal" weighting in mixed-data coefficients merely illustrates a general point: multimodal characters are to be preferred over unimodal ones and taxonomically correlated multimodal characters will have a stronger, clearer effect if they are redefined as binary or qualitative characters. Redefining quantitative characters as qualitative does of course produce some problems if there are borderline cases. Where the distribution curve is strongly modal but without gaps the dividing line between states should be placed at the dips. This minimizes but does not eliminate the cases of scores at or close to the dividing line being difficult to assign and introducing an element of error. If the qualitative character produced shows even some taxonomic correlation with others, then the conversion has probably been worthwhile. Certainly using this natural method of determining state definitions is superior to dividing the range into an arbitrary number of equal divisions.

In practice it is difficult to draw the frequency distribution curve if the taxa themselves are variable in their character. To produce histograms such as figs. 6.1 and 6.2, one must amalgamate measurements made on similarly sized and selected samples from each taxon. A better solution is to draw a bar diagram (after Tukey 1977), such as fig. 6.3, where each bar represents the range, the heavy bars span the 1st to 3rd quartiles, and the cross marks mark the median for the character in a taxon. Fig. 6.3 is for a morphological character in a study of *Narcissus* species by Almeida and Bisby (1984). It shows one clear discontinuity (marked a) and two slight breaks in the distribution (marked b and c). Notice that by placing the second state boundary at 20mm they were able to minimize the number of borderline cases. They contrasted this with a character that had no obvious breaks, in which case they had to resort to dividing the range arbitrarily.

Correlations Among Classificatory Characters

Correlation, the Basis of Classification

Correlations among characters produce the collective pattern used by taxonomists to make classifications. As explained in chapter 5, we mean by taxonomic correlation the association of discrete states or discrete parts of a quantitative scale and not the linear correlation often referred to by statisti-

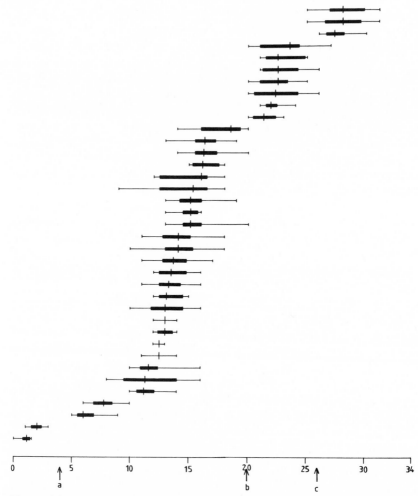

Figure 6.3 A bar diagram showing the distribution of samples from 37 taxa for a metric morphological character. Breaks in the distribution labeled a, b, and c could be interpreted as indicating four states of the character. *(From Almeida and Bisby 1984. Fig. 1).*

cians. These associations are seen because certain combinations of states recur among the items. So, for instance, if one examines the members of the Pea family (Leguminosae) there is a strong but not complete association between stamen number, sepal fusion, and petal arrangement so that of the 18,000 species in the family only 187 do not show one or another of the three character combinations tabulated in fig. 6.4.

Leguminosae (18,000 species)

Subfamily	Fruit type	Stamens	Petal arrangement at anthesis	Petal arrangement in bud	Sepal arrangement in bud
Papilionoideae	legume (pea pod)	10	descending imbricate *1	zygomorphic	fused
Caesalpinioideae	legume (pea pod)	8,10 or many	ascending imbricate	zygomorphic	mostly free
Mimosoideae	legume (pea pod)	8,10 or many	valvate *2	actinomorphic	united *3

exceptions: *1 185 spp. of tribe Swartzieae
          *2 1 sp. of Dinizia
          *3 1 sp. of Mimozyganthus

Figure 6.4 Table showing how correlated characters define three subfamilies of the Leguminosae.

The three combinations are used to delimit the three subfamilies (Papilionoideae, Caesalpinioideae, and Mimosoideae) and the only difficulty arises with the few species, such as those in the tribe Swartzieae, that do not fit the combinations exactly (Polhill 1981). Of course, very many other characters have been analyzed for the legume genera: some have proved of little use because they are correlated with others; and some show correlations within subsets of the family—that is, at other taxonomic ranks. None, however, is known to show such consistent correlation over the whole family as these three.

Character Analysis of Qualitative Data

The **character analysis** method based on information measure devised by Estabrook (1967) is useful for analyzing the correlation among binary and qualitative multistate characters, as well as quantitative ones divided into discrete ranges. His CHARANAL program performs the analysis on a taxonomic data table, such as the one given in fig. 5.13. The program takes characters two at a time and compares all possible pairs in turn. For each pair it calculates the total information content of the characters separately and the information held exclusively by each. From these two basic calculations a number of other information measures are calculated. The shared information—that is, the information held in common by the two characters—is directly related to the concept of taxonomic correlation and is of particular interest to us.

To define the information in common let us think of the two characters being compared as **a** and **b,** each deriving its information, H_a and H_b respectively, from its own way of partitioning the P items—that is, from the distribution of the P items over the states of characters **a** and **b.** Some subset of the items will be partitioned by the two characters in the same way. The information attributable to this subset is the information in common to the two characters which we shall designate $H_{c(a,b)}$ or simply H_c in fig. 6.5. The remaining subset of items will be partitioned differently by the two characters, producing two more portions of information—that portion in **a** but not shared by **b** and that portion in **b** which, likewise, is not held in common with **a.** These are designated $H_{a/b}$ and $H_{b/a}$ and are quantities defined in chapter 5 as "information held exclusively by **a** or **b**" respectively. They are calculated, as demonstrated in the last chapter, by summing over the information measures associated with the appropriate conditional probabilities.

There is an easy way to think about these various derived information measures using a representation known as a Venn diagram (fig. 6.5).

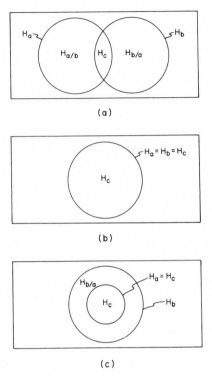

(a)

(b)

(c)

Figure 6.5 (a) A Venn diagram representing information held exclusively or shared by two characters a and b; (b) Venn diagram representing case where exactly the same information is held in two characters; (c) Venn diagram representing case where all information in a is also in b but the reverse is not true.

The information in each character, represented by the circles, is seen to be divided between that held exclusively by each and that shared between them. The total amount of information in each character is denoted by the relative sizes of the circles. The information in one or the other or in both characters, termed $H_{(a \cdot b)}$ in the last chapter, is the entire area within both circles. Further, the diagram suggests the simple relations:

$$H_a - H_{(a/b)} = H_{c(a,b)} = H_b - H_{(b/a)} \qquad (1)$$

and

$$H_{(a \cdot b)} = H_{c(a,b)} + H_{(a/b)} + H_{(b/a)} \qquad (2)$$

Thus, any of these four information measures can be found once the total information and the conditional measures have been calculated. The measures can then be combined into various ratios to give other derived measures. For example, the CHARANAL printout for each pair of characters includes the ratios $H_{c(a,b)}$ / H_a and $H_{c(a,b)}$ / H_b. These values give the information held in common as a fraction of the information in **a** and in **b** respectively.

A useful feature of CHARANAL is that it provides the equivalent of a correlation coefficient for qualitative data. For each pair of characters the program calculates a measure of independence known as D, standing for distance between two characters. It is calculated as:

$$D = \frac{H_{(a/b)} + H_{(b/a)}}{H_{(a \cdot b)}}$$

meaning that if the amount of information held exclusively in one or the other or in each of the two characters is large as a proportion of the total information in the characters, then D is large, i.e., near 1. Thus, the two characters are largely independent, that is to say, uncorrelated. By simply subtracting D from 1 we get a measure called S. S indicates the degree of correlation and is directly comparable to r, the common correlation coefficient for quantitative data mentioned in chapter 5. As is the case for r, values of S near 1 indicate correlated characters while values of S near 0 are associated with non-dependent characters. As is also true of r the question of exactly what S values indicate high or low correlation is a subjective one and the answer depends on experience with the measure. Generally in CHARANAL characters are considered to be highly correlated if S is over 0.3.

If two characters are perfectly correlated one would always predict the other: that is, if we knew which state of character **a** an object was in we could be certain of which state of character **b** it was in. If **a** and **b** have the same information content—that is, the same number of states and the same distribution of items in them, then this perfect correlation means that the two circles completely overlap in the Venn diagram and that $H_a = H_{c(a,b)} = H_b$. For example, if leaf shape, character **a**, and petal color, character **b**, divide some taxa in the same way as shown in fig. 6.5(b), then the information in both characters is equivalent and also equal. Another way to see this is to consider the simple relation (1) above and notice that since it is possible to predict states of petal color from leaf state and *vice versa* with certainty both $H_{a/b}$ and $H_{b/a}$ must be zero.

It is also possible for all of the information in, say, character **a** to be contained in character **b**, but not vice versa. In the Venn diagram for this case one circle is inside the other. This occurs when two or more states of a character can be nested within a state of a second character and is illustrated in fig. 6.5c. A practical example is that of the nested synapomorphies of Hennigian cladistics discussed in chapter 10, but a simple devised example is seen in table 6.2. One can calculate the information in **a** and **b** as:

$$H_a = -(\tfrac{5}{8} \log_2 \tfrac{5}{8} + \tfrac{3}{8} \log_2 \tfrac{3}{8}) = 0.954$$

and

$$H_b = -(\tfrac{1}{8} \log_2 \tfrac{1}{8} + \tfrac{4}{8} \log_2 \tfrac{4}{8} + \tfrac{3}{8} \log_2 \tfrac{3}{8}) = 1.405$$

We notice that while $H_{a/b} = 0$ the calculation for $H_{b/a}$ (not shown) gives 0.451, which is exactly equal to the difference between H_a and H_b. This is consistent with the representation of the situation in a Venn diagram as a pair of concentric circles, in which information quantity is represented by their relative sizes. It can also be seen from the first simple relation above that all the information in **a** is shared with **b,** but the reverse is not true.

Judgments about the usefulness of characters can be made on the basis of these pairwise comparisons alone or on an overview of the correlations of particular characters with all others. Judging on the pairwise comparisons, the taxonomist is interested in locating pairs of characters with relatively high values of information in common, $(H_{c(a,b)})$ which would be indicated by high values of S. These make a correspondingly large contribution to the collective pattern of the data set, as they are taxonomically correlated in many of the items. One should, however, be cautious of characters which are perfectly correlated: it is best to check both biological and logical aspects of the two characters. It occasionally happens that by accident two of the characters are in some way logically correlated by their definitions. An example might be:

Stipule presence: present   absent
No. of stipules: nil   1   2

where only after scoring is it observed for certain that the stipules are always absent or present in pairs, so that only two combinations are found: present and 2, or absent and nil.

Other complete correlations may depend on functional necessity such as the presence of a pollen-basket only on the legs of *Bombus* (Bumblebee) queens, which at certain times collect pollen, while the structure is absent in queens of *Psithyrus* (the very similar cuckoobee), which never collect pollen. In this case the two characters (Pollen Basket Presence, and Collect-

Table 6.2 An Example of Two Characters Having Nested Information

Plants	A	B	C	D	E	F	G	H
character **a:** presence of stipular nectaries	+	+	+	+	+	−	−	−
character **b:** petal color	R	B	B	B	B	W	W	W

ing Behavior) are completely redundant and should not both be included when the data are used for making a classification. However, where a correlation is marginally less than perfect, both characters should be retained. The occurrence of even one item in which they are not correlated establishes that the correlation is a biological one which is not a logical or functional necessity: that they can vary independently.

Further judgments can be made on taxonomic correlations by scanning the comparisons of, say, character **i** with each of the other characters. The justification for searching for correlated characters is explained using the information theory model as follows. The information in each character contains varying quantities shared with each of the other characters, and possibly some unique to itself. The collective pattern of prime interest to the taxonomist would surely match the small amount of information, if any, shared by all of the characters. For instance in the example of the three subfamilies of the Leguminosae already given (fig. 6.4), in addition to the four nearly perfectly associated characters already discussed, there are several other characters, which though strongly associated with the four, exhibit exceptions. As examples, the Mimosoideae subfamily contains nearly all plants with highly compound (2-pinnate) leaves but this character is varied in the other two groups, the Caesalpiniodeae are mostly trees or climbers, (as opposed to herbs and shrubs) and nearly all Papilonoideae have free rather than fused petals. Each of these characters, although imperfectly associated with the four characters, nonetheless shares some information with them: many pairs of species similar for one of these characters are also similar for the four; many species distinguished by one of these characters are distinguished by one of the four. A direct measure of this multishared information is not available, but what we can use as a second best is the sum of the fractions shared with each of the other characters. We assume that the character with the largest sum of fractions shared with others is actually the character that contributes most to the multishared information. The sum,

$$\sum_{j=1}^{n} \frac{H_{c(i, j)}}{H_i}$$

is calculated by the character analysis program CHARANAL (where it is called SAMRAT) once all pairs of characters have been compared.

Table 6.1, above, is taken from a study of 273 species of African *Crotalaria* in Bisby (1970), and shows the characters used, listed in order of descending information contribution calculated as shown above. An interesting feature of the results is the position of characters marked *PP, p,* and Ba. Those marked *PP* and *p* had recently been used by a taxonomist employing only traditional methods (Polhill 1968) to classify the genus into sections and subsections, those marked *PP* for sections, *p* for subsections.

A previous classification, rejected as unsatisfactory, was by Baker (1914) and used for the major divisions the characters labeled *Ba*. The analysis supports Polhill's contention that his sectional and subsectional characters are better than those of Baker: all of the characters marked *PP* come above those marked *Ba* in the list.

Character Analysis of Quantitative Data

Techniques for calculating the linear correlation between pairs of characters and between one character and all others for continuous data are available. However as discussed in chapter 5, there is some question as to how appropriate they are for taxonomic use. Let us look at some specific examples here.

Consider two well-correlated characters that are qualitative as in the Leguminosae characters given above:

> Petals: zygomorphic or actinomorphic
> and
> Petals: imbricate or valvate

These two characters are strongly correlated in 18,000 species and are of high taxonomic value in supporting the position of subfamily Mimosoideae separate from the remainder. Now consider two well-correlated metric characters such as ovary length and keel length in *Vicia sativa* (Hollings and Stace 1978). As shown in fig. 6.6 there is a high linear correlation between

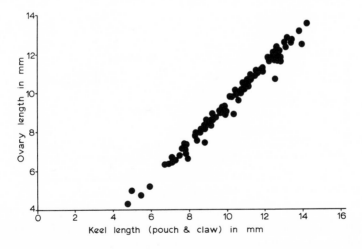

Figure 6.6 An example of a high linear correlation between the two characters, Keel Length and Ovary Length in a vetch. *(From Hollings and Stace 1978. Fig. 8).*

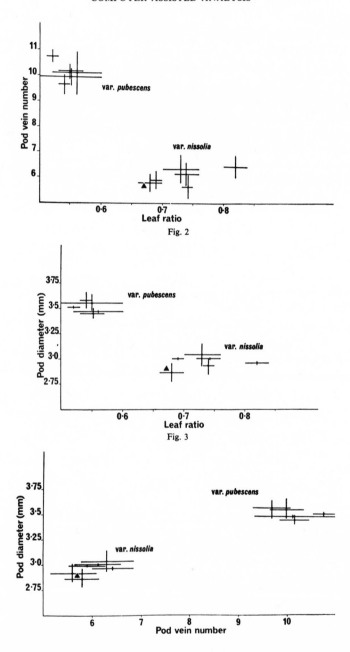

Figure 6.7 Diagram illustrates taxonomically useful correlation in that there are discontinuous groupings based on all possible pairs of the three characters. *(From Cannon 1964. Fig. 4).*

the two characters as measured by the product-moment correlation coefficient. Despite this they are of no use taxonomically in this context because taken together they fail to produce discontinuous groupings. Fig. 6.6 was in fact given by Hollings and Stace to explain why they rejected these characters. In contrast the quantitative characters diagrammed in figs. 6.7 and 6.8 do produce discontinuous groupings, and hence would be considered to be usefully correlated characters by taxonomists, although they do not have a high linear correlation. Notice that even though these are quantitative characters their distributions are discontinuous and they could be treated as multistate disordered characters. When so treated, states of the two characters are associated or correlated in the sense detected by the information measure.

Figures 6.7 and 6.8 show different ways of diagramming so as to detect patterns of taxonomic correlation among characters. In fig. 6.7 Cannon (1964) has drawn crosses to represent samples for which the lengths and positions of the vertical and horizontal bars represent the sample means ± one standard error. The scatter diagrams show how the three characters are taxonomically correlated in the sense already described. As each of the characters appears to have a clearly bimodal distribution, it should be possible to convert the characters to three binary characters which would then be perfectly associated—that is, with all their information in common. Fig. 6.8 from Agnew (1968) shows a similar example except that one of the

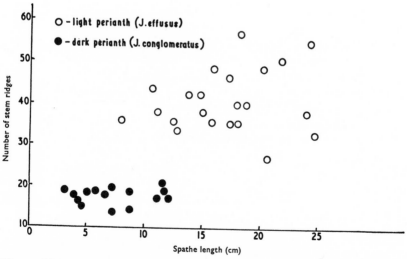

Figure 6.8 Diagram of clusters separated on the basis of three characters. *(From Agnew 1968. Fig. 1).*

characters is a binary character and that the data are recorded for individuals, not samples.

When two variables are linearly correlated the points plotted on a scatter diagram lie on a line. Therefore, Principal Components Analysis (PCA, see chapter 5) detects pairs or sets of highly correlated characters when detecting principal axes. Those characters with high positive loadings on a particular axis are highly (linearly) correlated. Those with high negative loadings are linearly correlated with each other too, but negatively correlated with those with positive loadings. However because of our strong reservations about the taxonomic significance of linear correlations we see little taxonomic application for these results.

Character Selection and Weighting

Although taxonomists select and weight characters frequently, this activity has received less than its share of attention from taxonomic theoreticians and mathematicians, and rather more than its share of abuse from extremist advocates of the Adansonian method.

Selecting characters and weighting characters are variants of the same logical process of choosing what relative weight to give different characters. If certain characters are selected for inclusion in an unweighted analysis, these characters are given relative weights of 1.0 and others (those rejected) are given relative weights of 0.0. Alternatively, taxonomists who weight characters, for whatever reasons, assign to selected characters varying weights (say 0.5, 1.0, 1.5), and again reject other characters by weighting them 0.0. Some authors have claimed to use all available characters but however this is defined there will be some that are rejected so that effectively selection or weighting has occurred.

Before the widespread acceptance of natural classifications, botanists decided before making a classification what characters were going to be given prominence. This is the so-called *a priori* selection of characters for reasons either of convenience or of dogma. Aristotle's division of plants by their uses, or a nurseryman's separation of annuals, perennial herbs, and shrubs, are examples of the former; Linnaeus' use of stamen number or Hutchinson's adherence to woodiness characters are examples of the latter. The vehemence of many Adansonian extremists is directed not so much in favor of giving many characters equal weights of 1.0 after Adanson's *method de l'ensemble* as against this *a priori* selection and the artificial results it produces. *A priori* weighting and Adansonian unweighted analyses are not, however, the only possibilities. There are a variety of *a posteriori* methods that analyze the value of characters after they have been scored.

Just as we have already discussed the value of characters in terms of intrinsic properties of a single character and of taxonomic correlations between them, so too can these be used as reasons for weighting them. One can think of these methods either as highlighting weak correlations or collective patterns in a data set, or as playing down uncorrelated or unpatterned elements of the data. Some taxonomists (see for instance Davis and Heywood [1963]) think that such activities are at the heart of good traditional taxonomic practice.

Adams (1975) compared six methods of giving individual quantitative characters weights related to an *a posteriori* analysis of their intrinsic properties in samples from species of *Juniperus* to be classified. Some of these methods gave weight to characters that varied widely within samples and others gave increased weight to those that varied much between samples but little within samples. His results show a scale of effects, with weighting in proportion to a character's F ratio (variance between samples/variance within samples) giving the most meaningful and no weighting at all giving the least meaningful results. It should be pointed out, however, that he was working with crude characters—chromatograph peaks taken from a survey of terpenoids. The equivalent of separating meaningful peaks that varied principally between samples from meaningless peaks that varied as well within samples would, in other kinds of data, probably have been performed subconsciously by a traditional taxonomist. Whiffin (1982) has used the same technique for volatile oil data on trees in tropical rain forests.

Goodall (1966) and Burnaby (1970) have proposed methods which also weight individual characters but on the basis of the rarity of their states. They have been little used. As stated earlier, the value of characters with rare states in our opinion cannot be evaluated for individual characters: it lies in the extent of taxonomic correlations of such characters with others.

Weighting or selecting characters according to a measure of taxonomic correlation has also been carried out rather rarely. Examples of selections based on the information contribution from Estabrook's character analysis are given for African *Crotalaria* by Bisby (1970) and for the *Cytisus/Genista* complex in Bisby and Nicholls (1977). Bisby selected the 36 best correlated characters from a set of 48. The information contribution values of all 48 characters were listed and a cutoff point chosen at which the values had become relatively uniformly low. Experience shows that the values form a curve which flattens out, but the cutoff point chosen is arbitrary. In the *Cytisus/Genista* complex study, which was performed to test for stability of character sets, Bisby and Nicholls chose the better-correlated half of the characters. In both studies the selection led to a clearer pattern and one which gave better agreement with conventional results. The method selects rare-state characters only if the rare-state is well correlated with other rare-state characters.

Further Development of Character Analysis

We conclude that in taximetric as in conventional studies taxonomists must give much attention to the choice and formulation of characters. Characters are the medium through which variation is surveyed, recorded, analyzed, and communicated. No matter how sophisticated a taxonomic study is, it can always be faulted if the characters are poorly structured or poorly observed. Authors of early numerical taxonomic publications, who gave the impression that the choice and formulation of characters was a single operation, do themselves a disservice. In practice, in computerized studies characters undergo the same scrutiny, the same gradual evolution, the same alteration for use as different types of sister characters, as do characters in conventional studies.

A taxonomic character derives its classificatory value from the distribution of its character states as well as the taxonomic correlation of this distribution with the distributions of other characters. Information measures used in character analysis can be of value in choosing, making improvements, or weighting characters after they have been recorded, but prior to their use in pattern analysis or classification methods such as those described in the next chapter.

Notwithstanding what is written down, our knowledge of both the biological and the mathematical aspects of character selection and weighting is woefully slender. We have not come far enough in dispelling the comment of Davis and Heywood (1963), writing of numerical methods, that "Character selection is the weak link in the whole approach." We disagree only in that the weak link is not peculiar to numerical methods; indeed if the matter were clarified for conventional methods it would probably be relatively simple to incorporate the clarification into numerical methods.

Suggested Readings

General

Davis, P. H. and V. H. Heywood. 1963. *Principles of Angiosperm Taxonomy*. Edinburgh and London: Oliver and Boyd.

Sneath, P.H.A. and R. R. Sokal. 1973. *Numerical Taxonomy*. San Francisco: W. H. Freeman.

Discussion

Allkin, R. 1984. Handling taxonomic descriptions by computer. In R. Allkin and F. A. Bisby eds. *Databases in Systematics*, pp. 263–278. London and New York: Academic Press.

Pankhurst, R. J. 1984. On the description of inflorescences. In Allkin and Bisby eds., pp. 309–320.

Almeida M. T. and F. A. Bisby. 1984. A simple method for establishing taxonomic characters from measurement data. *Taxon* 33:405–409.

Chapter 7

Phenetic Classification

The methods described in this chapter are used to analyze the phenetic variation patterns among organisms. The results provide taxonomists with the basis for new or changed classifications. Some of the methods provide an analysis ready-made in the form of a classification, others need considerable interpretation. Phenetic patterns and phenetic classifications are based on overall resemblances among organisms for characters assumed to be heritable.

The practice of computer-assisted classification may seem to involve no more than putting the data, in machine-readable form, into a computer that is programmed to analyze or classify it automatically. However, for us to stop with such a superficial description or even simply to list the names of suitable computer programs would be irresponsible. The essence of computer-assisted classification lies in the computer program's ability to carry out an explicit pattern analysis and classification procedure—a procedure that is precise and repeatable. The strength, or as some see it the weakness, of the exercise is that the user must choose the exact procedure to be used.

In every case the procedures involve simple logical or arithmetical steps which could, at least for small data sets, be carried out with paper and pencil or a calculator. However, the number of calculations involved usually rises sharply with the quantity of data, so that time is saved and accuracy improved if a computer is used. The reader should not be deceived by claims that a data set was analyzed by a computer in a fraction of a second! Such claims usually refer to the time for calculations within the computer and don't count the time used for input, output, and routine file management. At a large computer installation it would be rare for the results to be available in less than 30 minutes of supplying the data to the machine. When time for preparation of data and comprehension of the results is considered, there may well be a week's work or more.

Conceptually the procedures for computer assisted classification are derived by analogy from the mathematical models described in chapter 5. The variety of models, and consequently of procedures, reflects the many conceptions of how taxonomists analyze variation patterns and make classifications. Occasionally, as in the case of single-link cluster analysis, the same procedure has been derived independently several times from different models. All the methods are multivariate in that they analyze the collective pattern based on resemblance among many characters. We can call these patterns "natural patterns" in the neo-Adansonian sense of natural resemblances being manifested as overall similarities.

Ordination techniques are used to provide a visually displayed simplification of the variation pattern. The **cluster analysis** techniques go a step further by analyzing the variation pattern and constructing a hierarchical classification based on it. It is these cluster analysis techniques which have proved so useful to taxonomists and which will receive most of our attention.

Agglomerative Two-Step Cluster Analysis

The cluster analysis methods most commonly used in taxonomy are the agglomerative two-step ones, as shown diagrammatically in fig. 7.1, consisting of:

1. Constructing a **similarity** or **distance table,** containing values for the overall resemblance between all possible pairs of items.
2. Constructing diagrams using a **clustering procedure.**

There is a reasonable amount of freedom to combine any appropriate first step with an independently chosen second step, so we shall treat the two separately. What is of paramount importance for the first step is that the measure of resemblance exactly matches the form and meaning of the data. In the second step clusters are recognized according to precise criteria. The user should select the criteria which are needed and then use the appropriate clustering method.

Resemblance Measures

If one starts with a tabular data structure, as shown in chapter 4, then the first step is to calculate the resemblances for each item compared with every other item. There is a wide range of measures of resemblance, some of which

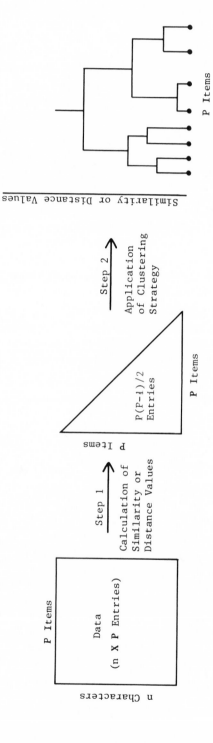

Figure 7.1 Diagram of overall procedure in agglomerative two-step cluster analysis.

are discussed below. All may be thought of either as measures of similarity (in which the higher the value the more similar the items), or of distance (in which the higher the value the less similar the items). The table is, correspondingly, known as the similarity matrix or the distance matrix. Resemblance measures are usually recorded in the form of a triangular table, because values in one half of the full square matrix mirror the values in the other half, since the resemblance between objects **A** and **B** is equal to that between **B** and **A**. The row of entries along the diagonal contains the resemblances of each item to itself, ordinarily equal to 0 for distance measures and to 1 for similarity measures.

The most important features that must be taken into account when choosing a measure of resemblance are the exact type of data presented and the kind of resemblance desired by the user. We will now describe a number of specific measures in terms of these criteria.

Measures for Quantitative Data
The **Euclidean distance** measure is the distance between two points in the geometric model, calculated from Pythagoras' theorem extended to n dimensions as explained in chapter 5 and summarized in the following formula:

$$Di_{AB} = \sqrt{(a_A - a_B)^2 + (b_A - b_B)^2 + \cdots + (i_A - i_B)^2 \cdots + (n_A - n_B)^2}$$

$$= \sqrt{\sum_{i=a}^{n} (i_A - i_B)^2}$$

Absolute values of this measure rise with the number of characters. If one wishes to make comparisons of values between different studies having different numbers of characters, then an average difference, Di_{AB}/n, is calculated where n is the number of characters. Also for simplicity of calculation Di_{AB}^2 is sometimes used.

The closely related **Manhattan block distance** utilizes the absolute distances in the geometric model from **A** to **B** along lines parallel to the axes (i.e., excluding diagonals as in city blocks). Thus:

$$Di_{AB} = |a_A - a_B| + |b_A - b_B| + \cdots + |i_A - i_B| \cdots + |n_A - n_B|$$

$$= \sum_{i=a}^{n} |i_A - i_B|$$

These four measures are the most frequently used means of measuring overall resemblance among items for which quantitative numeric and ordinal data are recorded. If the distances along the axes in the geometric model are thought of as numbers of mutations, then the Manhattan block distance becomes a mutation distance. It must be remembered that the value of the measure used lies in the goodness of the analogy between it and the taxonomist's notion of dissimilarity. Thus, for instance, Euclidean distance is not always preferred in taxonomy even though it measures the shortest distance between two points in physical space.

The **product moment correlation coefficient,** r_{AB}, can be used as a measure of similarity between two items **A** and **B** although it has a number of peculiarities when used in this way. It measures the extent of a direct or inverse linear correlation between all the character values of items **A** and **B,** but not the similarity in their magnitudes. The maximum value of $+1$ does not mean that **A** and **B** have identical data as shown in chapter 5 and the minimum value of -1 has no obvious biological meaning and is very inconvenient computationally. Where users have a choice our general recommendation is that they avoid this measure. The one case where it does seem sensible is when one has data that are entirely quantitative comparisons of the sizes of various body parts, so-called morphometric measurements. Here the correlation measure can successfully compare the shapes of two objects, **A** and **B,** without being affected by differences in size alone.

Measures for Qualitative Data

We now look at measures suitable for measuring the resemblance between two items **A** and **B** scored for a number of qualitative characters. If the qualitative characters are binary with possibilities scored as plus or minus, then there are a number of measures most easily described using a contingency table. In the notation of the contingency table in table 7.1, the lowercase a, for example, means the actual number of characters scored with the positive state for both item **A** and item **B**. The lowercase b means the actual number of characters scored negative for **A** and positive for **B**, and so on for c and d. The sums indicated in the margins are row or column totals that add up to n, the total number of characters.

Table 7.1 A Contingency Table

		Item A		
		No. characters scored +	No. characters scored −	
Item B	+	a	b	$a+b$
	−	c	d	$c+d$
		$a+c$	$b+d$	$a+b+c+d=n$

The **simple matching coefficient** is a measure of similarity calculated as Ma_{AB}, the number of data matches between **A** and **B** divided by n, the number of characters. Thus,

$$Si_{AB} = \frac{Ma_{AB}}{n} = \frac{a+d}{n}$$

The complement of Si, calculated as $(1 - Si)$ is a distance Mm_{AB}, the number of mismatches between **A** and **B** divided by n. Thus,

$$Di_{AB} = \frac{Mm_{AB}}{n} = \frac{b+c}{n}$$

Both of these measures are used for symmetric binary data: they count the a double positives as of equal value to the d double negatives. In cases where the user wishes to ignore double negatives, that is to treat binary characters as asymmetric, a **nonmetric coefficient** (Jaccard 1908, as cited in Sneath and Sokal 1973) is used as in the following:

$$Si_{AB} = \frac{a}{a+b+c}$$

Two fairly common uses are in bacteriology, where the characters are responses to tests in which only a positive response is thought to be informative, and in scoring secondary chemical data in flowering plants, where the negative score means the substance was unreported possibly because no one had examined that plant for that chemical.

All three measures can be used for multistate disordered characters if expressed in terms of matches or mismatches. Users with multistate ordered characters have a choice. If all the characters are of this form and the user wishes each to be treated on a linear scale, the measures for quantitative data given above should be used. Alternatively, departures from a linear ordering or partial ordering can be achieved with the mixed data coefficients given below.

Measures for Mixed Data

In general, the various similarity measures that have been devised for mixed quantitative and qualitative data are not too different. They begin from the principles of the simple matching coefficient applied to strictly qualitative data. Each matching state of a single character, when one considers a particular pair of objects, contributes a maximum of 1 to the resemblance measure. Each mismatch for simple qualitative data contributes 0 (assuming symmetric character states in the binary case). Quantitative or ordered multistate data contribute an amount to the similarity measure between 0 and 1 according to a preset rule for assignment. Once the contributions, Co_i, for each character—of whatever type—are determined, the similarity between two objects is determined by summing. Thus,

$$Si_{AB} = \sum_{i=1}^{n} Co_i$$

Mixed data coefficients differ according to the rules used to assign the contribution for the various types of data. The rules of Gower (1971) and Estabrook and Rogers (1966) are shown below.

Gower's mixed data coefficient
For binary data (assumed to be asymmetric):
  $Co_i = 1$ for a double positive match and 0 for a mismatch or a double negative match.
For simple multistate data:
  $Co_i = 1$ for a match and 0 for any mismatch.
For quantitative or ordered multistate data:
  $Co_i = x$, where x is between 0 and 1 and is determined as $x = i_A - i_B$ after scaling the character i on its range (see chapter 4).

Rogers and Estabrook's mixed data coefficient
For both the forms of qualitative data, binary (assumed symmetric) and simple multistate, $Co_i = 1$ for a match and 0 for a mismatch.
For a k state ordered multistate character or an ordinal character, Co_i is computed as the average difference between the ordering numbers of the states, computed on $k - 1$ states. For a fully quantitative or nonlinearly ordered multistate, the differences between all possible states of each character must be specified in a character state matrix.

Cluster Strategies

We shall start this section by describing single-link clustering in some detail. Readers should be sure that they can follow the steps and work through the examples. Single-link clustering is an important model in its own right; furthermore, several other clustering methods can be thought of as variants with broadly similar steps. Single-link clustering is an algorithm that has been "discovered" several times, derived from different mathematical models and, confusingly, given different names. We give two derivations below: The first employs set and graph theories and yields linkage diagrams that are of great use to taxonomists. The second single-linkage, or nearest neighbor, procedure uses the geometric model and results in dendrograms which, though less useful in some ways, do provide a means of comparison with other methods also derived from the geometric model. Single-link clustering was first published as the "dendritic method" by Florek and coworkers in Poland (1951 a and b).

Single-Link Cluster Analysis

Estabrook (1966) derives the single-link method from an axiomatic set theory model; he decides first on some properties which will be axiomatic or assumed. Reworded, his axioms are as follows:

1. A biologically meaningful measure with metric properties is available to measure similarity between items.
2. A biological classification for a collection of items is a nested hierarchical series of partitionings of the items into classes.
3. At a given threshold of similarity, two items that are more similar than the threshold should be placed in the same class, thus forming one of the partitions. (The method is therefore phenetic, depending on similarity, not descent.)
4. The classes of a given partition should be isolated from one another; that is, there should be some discontinuity in similarity between the members of different classes.

It can be shown that Axiom 2, although correct, can be deduced from 3 and 4, and that Axiom 4 is to some extent the reverse (contrapositive) of 3. The minimum needed is to retain Axiom 1 and substitute the following for numbers 2, 3, and 4. We shall call this the single-link axiom.

An item is a member of a class if and only if it has a similarity value greater than a given threshold similarity with at least one member of that class.

Clearly the procedure implied by these axioms gives rise to classes. Items with similarities to all other items less than the threshold will be in singletons—that is, classes with only one member. By raising or lowering the threshold different partitions will be formed. When the threshold is raised progressively, it can be demonstrated that the partitionings form a series of hierarchically nested classes as prescribed in the original Axiom 2.

Estabrook points out that series of clusters at any level can be portrayed in diagrams he calls **subgraphs** (from graph theory). We shall henceforth call these subgraphs **linkage diagrams.** For any threshold similarity level c, we can represent the items as points in arbitrary positions and use the line to join only those items whose similarity exceeds c, as shown in fig. 7.2.

A second derivation of single-linkage, this time from the geometric model, looks very different but in fact arrives at the same computations. We start this time with a distance table and think of the distances as geometric distances between points representing items in a multidimensional space. (See chapter 5 for details.) The cluster procedure is a cyclic process based on locating first the closest two points in the space, then the next closest pair, then the next, and so on. We can think of this repeated process as

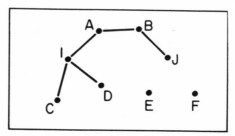

Figure 7.2 Example of a graph containing a sub-graph or linkage group.

a visual search in the space, but the computer program does it simply by searching the distance table for the next smallest value. (If similarity values are used instead of distance values, then the search must be for the next highest instead of the next lowest value.) Whenever this next value is located there are four possible conditions that can arise. Each of these cases must be treated slightly differently, as we describe below. [In the diagrams of the single-linkage examples in figures 7.3 through 7.7 the parts of the figures labeled (a) are linkage diagrams and those labeled (b) are dendrograms.]

Case 1: The items located by the threshold criterion have been so far unclustered and each pair is now declared to form a new cluster. So, for instance, the two items **B** and **F** in fig. 7.3(a), separated by the shortest distance 0.4 from each other, are fused or linked to form cluster **BF.** We keep track of the fusion by drawing it on a hierarchical diagram called a dendrogram. The crossbar or bracket joining **B** and **F** is drawn at level 0.4 on the vertical distance scale as shown in fig. 7.3(b).

This procedure is repeated, possibly producing several two-member clusters like **BF,** but at progressively higher levels. If there are two or more pairs of points at the same distance apart, they are taken in arbitrary order, but of course produce clusters at the same level. Alternatively, if the two pairs have one item in both pairs, as in **EG** and **GS** in fig. 7.4, the three items form a three-member cluster.

Case 2: At some stage, depending on the arrangement of the

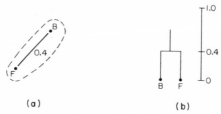

(a)   (b)

Figure 7.3 Case 1. Diagram of the linking of two items with dendrogram showing the two items clustered at $c = 0.4$.

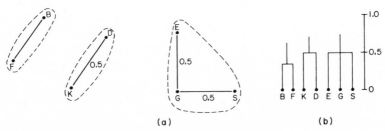

(a)   (b)

Figure 7.4 Formation of a three-member cluster at $c = 0.5$.

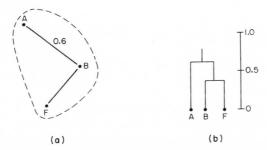

(a)   (b)

Figure 7.5 Case 2. Linking an item with a previously formed cluster at $c = 0.6$.

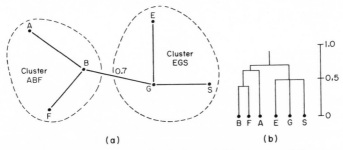

(a)   (b)

Figure 7.6 Case 3. Linking two previously formed clusters at $c = 0.7$.

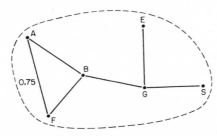

Figure 7.7 Case 4. Formation of an internal link at $c = 0.75$.

points, the next closest pair causing the next link will include one point which is already a member of a cluster. When this occurs, the single link criterion is used—that is, the item is fused with the cluster at the level of the first link with one of its members. (This is in contrast to other methods, where more than one is required.) See fig. 7.5.

Case 3: Similarly, sooner or later the next closest pair causing the next link will include one item in one cluster and one in another, say **B** in **ABF** and **G** in **EGS** in fig. 7.6 Again the single-link criterion fuses them at the level of the link (**BG**) first linking a member of one cluster with a member of the other.

Case 4: Eventually the case will occur where the next closest pair of items forming the next link are both members of the same cluster. This is called an **internal link,** and is ignored for the purpose of drawing the dendrogram, since it makes no change in the membership of clusters. This is in contrast to the linkage diagram method where internal links are shown. See fig. 7.7.

The cyclic process of examining next-closest pairs continues, each pair located falling into one of the cases 1 to 4 above. Eventually all the items are fused into one large cluster and the analysis is complete. The reader will be able to see from the diagrams in fig. 7.8 that the method does indeed detect clusters of points in the space. However it should be noted that in the case of **MLDK** it has detected a long, thin cluster in which **K** is some distance from **M**, whereas **BFAEGS** is more of a dense cluster.

Single-linkage cluster analysis has two properties of considerable importance for taxonomists. First, clusters are effectively delimited in the geometric model by gaps around them, or in the axiomatic set theory model by similarities below the threshold, such a gap being referred to as the moat by Estabrook. The analogy for a taxonomist is that taxa are delimited by discontinuities. If we take **LMDK** and **BFAEGS** in our example, we know that no link occurs between the clusters until level 0.95. Thus, there is a moat of 0.95 surrounding the two clusters. The size of this moat can be

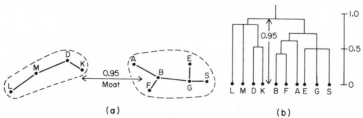

Figure 7.8 Dendrogram of a long, thin cluster (**MLDK**) compared with a more dense, compact cluster (**BFAEGS**); note moat of 0.95.

read directly from the dendrogram. With linkage diagrams the moat is indicated by the similarity value for the diagram in which the clusters first fuse with others. Second, when an item joins a cluster, it always joins the same cluster as the item to which it is closest. In taxonomic terms this means that a taxon is always grouped with the taxon to which it is most similar.

It is interesting to take as examples of applications of single-linkage cluster analysis the work of Watson, Williams, and Lance (1966, 1967) on the classification of Epacridaceae using dendrograms and the work of Bisby and Polhill (1973) on classifying African *Crotalaria* (rattle pods) using linkage diagrams. Watson et al. used a single-linkage cluster analysis ("'nearest neighbour" as they called it) for 20 qualitative binary characters scored on exemplar plants from 24 genera of the flowering plant family Epacridaceae. Genus 24 was *Wittsteinia,* a genus which various orthodox taxonomists had classified in different families—in the Vacciniaceae (Bentham in 1879), in the Ericaceae (Drude in 1889) and finally in the Epacridaceae (Burtt in 1948). The results reported were very similar for the single-linkage analysis and one other method (the true centroid method of analysis which we shall discuss later in this chapter). *Wittsteinia* (and another genus *Needhamia*) were isolated and the others generally formed two major clusters which corresponded with the tribes Epacrideae and Styphelieae of Bentham's system.

Watson and his coworkers were well impressed with the stability of the groupings produced, but we now know that with data having a less obvious pattern this is not always so. Indeed, in their next paper Watson et al. (1967) reject single linkage in favor of true centroid clustering with an argument that has been followed by many. They state that "A most important requirement in any primary taxonomic survey is the delimitation of clear-cut groups for further study: and we already have evidence that 'centroid' sorting is more efficient in this respect than 'nearest neighbor' sorting."

We do not hold that opinion: indeed the reverse argument is more persuasive. In a primary taxonomic survey it is important to display the full variety of elements in the variation pattern so that homogeneous and heterogeneous clusters, outlying and intermediate taxa, chains of intermediates and chains of outliers can all be spotted so that decisions on classifying them can be taken "in full possession of the facts." Bisby and Polhill provide linkage diagrams for 273 species of African *Crotalaria* as shown in fig. 7.9.

They found that the most useful diagrams were those at levels not where clusters were distinct, but at which clusters link, or in some cases chain, together. Fig. 7.9, for instance, is the first level at which the major dense clusters (shown as circles) link together, in many cases via intermediates. To facilitate comparisons with an earlier orthodox classification by Polhill, the species are represented by symbols showing to which sections of

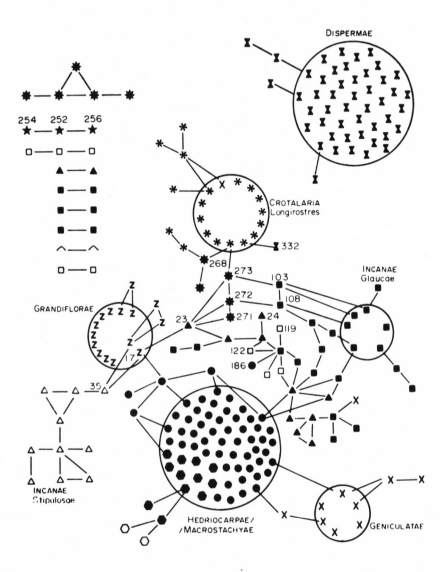

Figure 7.9 A linkage diagram for the analysis of relationships among species of *Crotalaria*. *(From Bisby and Polhill 1973. Fig. 1).*

Polhill's 1968 system they belong. For convenience only intermediate or linking species in interesting positions are identified by number on the diagram. Complicated as this diagram may seem, it does contain very much of the pattern information, which we think is useful to a taxonomist. The chart's depiction of Dispermae, for example, is that of a large homogeneous group with few outliers and in a more isolated position than other major groups. Two subsections of Incanae (solid squares and triangles) clearly constitute a heterogeneous assemblage bridging between other clusters. Grandiflorae and subsection Stipulosae are both accurately distinguished in the diagram, but we can see that the Grandiflorae, despite crosslinks with other clusters, is more homogeneous than the isolated but diffuse Stipulosae. Bisby and Polhill obtained such an accurate correspondence between the linkage diagram and Polhill's orthodox views of the genus that they were able to list the few disparities and find reasons for each, such as erroneous data-scoring by Bisby or erroneous observation by Polhill. The accuracy of this correspondence serves to give considerable confidence to the users of this technique.

Group-Average Cluster Analysis

This method differs from single-linkage in that the criterion distance for a candidate item or an established cluster to join a cluster is not a single distance, but the group-average distance—that is, the arithmetic mean of all distances between the candidate(s) and the items in the cluster. The name used here is an alternative for that used by Sokal and Michener, the "unweighted pair group method using arithmetic averages" (abbreviated to UPGMA) when they introduced the method in 1958. (We prefer to avoid the use of 'weighted' and 'unweighted' in the names of this and following methods because doubts as to what is or is not being weighted have occasionally led to confusion in the literature.) A detailed worked example of group-average clustering is given in Sneath and Sokal (1973).

The group-average cluster strategy can be thought of as a geometric process following broadly similar lines to the geometric version of single linkage given above. As before distance measures between all pairs of items are stored in a triangular distance table, and are thought of as interpoint distances in a multidimensional geometric model. As before the closest pair of points and then each successive, closest pair in the model are detected by locating the lowest and the next lowest, etc., values in the table. The equivalent of cases 1, 2, and 3 of single linkage occur again. Case 1 is identical: the two items located as having the next smallest interpoint distance are fused to form a cluster.

Cases 2 and 3 differ. When a cluster is formed group-average distances are calculated between it and other items and other clusters. These values are added to the distance table and distances from individual cluster members are deleted. This means that in case 2, the fusion of an item with a cluster, the fusion occurs when a value such as $Di_{D,ABC}$ in fig. 7.10(a) is the next lowest value in the table. Case 3, the fusion of two clusters, occurs when a value such as $Di_{FD,ABC}$ in 10(b) is the next lowest in the recalculated table. It should be noted that in these two examples, as in most cases, the group-average distances for fusions 2 and 3 are considerably larger than in the single-link distances, which would have been Di_{DA} in (a) and Di_{FA} in (b). So the result is that the crossbars in a dendrogram are slightly higher, or alternatively that items and clusters, or two clusters, must be closer together to fuse at a given level on the dendrogram.

   Group-average cluster analysis is often chosen because of its supposedly intermediate characteristics, lacking the extreme chaining property of single-linkage, and at the same time lacking the tendency to exaggerate groups of the central point methods. Others argue that it comprises the "worst of both worlds," as it is lacking both the precision of the single-link axiom and the "powerful" tendency of space-dilating methods (see below) to emphasize gaps and form a balanced hierarchy. McNeill (1975) uses the group-average method as one of six cluster analyses whose performances he compares in classifying a tribe of the flowering plant family Portulacaceae. His dendrograms, given in fig. 7.11, show that the group-average method is in fact much nearer the single-link end of the scale. The dendrogram distinguishes a dense cluster of the species nos. 1–24 to the left of the dendrogram, the genus *Claytonia sensu* Swanson, from the other more widely separated species which McNeill, perhaps with a little less confidence, groups as a single genus *Montia*.

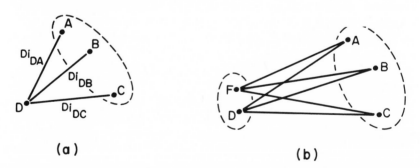

Figure 7.10 (a) Case 2. The fusion of an item with a cluster is based on the average distance of the item from the cluster members calculated as $Di_{D,ABC} = (Di_{AD} + Di_{BD} + Di_{CD})/3$. (b) Case 3. The fusion of two clusters, which is based on the average distance between all possible pairings of members of the two groups, calculated as $Di_{FD,ABC} = (Di_{AD} + Di_{BD} + Di_{CD} + Di_{AF} + Di_{BF} + Di_{CF})/6$.

Central Point Cluster Methods

Here we consider two cluster analysis methods in which clusters in the geometric model are replaced by their central points—**median cluster analysis** and the **true centroid method.** The methods differ in the detail of where the central point is placed. In both cases the criterion for fusions of a cluster to items or to other clusters is the distance from this central point and not, as in single-linkage, from the nearest cluster member nor, as in group average, from all of the cluster members.

The procedure in central point clustering follows a sequence like that in single-linkage and group-average methods. A distance table is thought of as a multidimensional geometric model and the table searched for the lowest and next lowest values, etc. Again in case 1, when the next lowest distance is between two items as yet unclustered, they are fused to form a cluster, and the distance between them is the height of a crossbar on the dendrogram. Once a cluster is formed, individual distances from its members are deleted and replaced by distances from the cluster's central point to other unclustered items or central points. In case 2, the fusion of an unclustered item and a cluster, and case 3, the fusion of two clusters, the criterion distances are those to the central point. Where clusters have only two members those distances will be as depicted in fig. 7.12.

The two central point methods differ in the placement of the central points in clusters with three or more items. Imagine that a cluster containing three members, **A,B,** and **C,** are formed by the fusion of a two-member cluster, **A,B,** with its central point at **X,** and the unclustered point **C,** as in fig. 7.13(a). In the median cluster analysis method the new central point will be placed halfway along the line **CX**—that is, halfway between the centers of the two cluster representations to be fused, **C** and **X,** fig. 7.13(a). It can be argued that this gives the two subclusters equal weight or alternatively that the constituent members are weighted, with **C** getting twice the weight of **A** and **B.** Similarly, when two clusters are fused, the new central point is placed at the midpoint of the line joining the central points of the subclusters [see fig. 7.13(b)]. Confusingly, this method is also known as the Weighted Pair-Group Centroid Method, WPGMC.

In the true centroid method the new central point is placed at the true centroid (multidimensional center) of all the points in the cluster. In our example of a four-member cluster joining an item, the new centroid is placed on the line **CX** but nearer the **X** end, so that the line is divided into two pieces of relative length 1:4, that is, the ratio of numbers of items in the clusters being fused (see fig. 7.14). It can be argued that this gives one subcluster greater weight or alternatively that the individual items receive equal weight. It is sometimes known as the Unweighted Pair-Group Centroid Method (UPGMC).

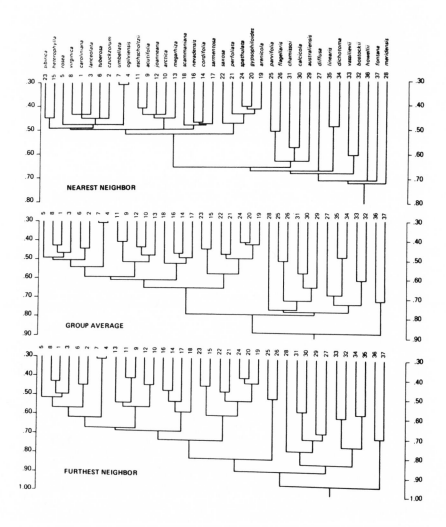

Figure 7.11 McNeill's dendrograms resulting from six different forms of cluster analysis applied to the same data for the tribe Portulacaceae. *(From McNeill 1975. Fig. 2(a) and 1(b).*

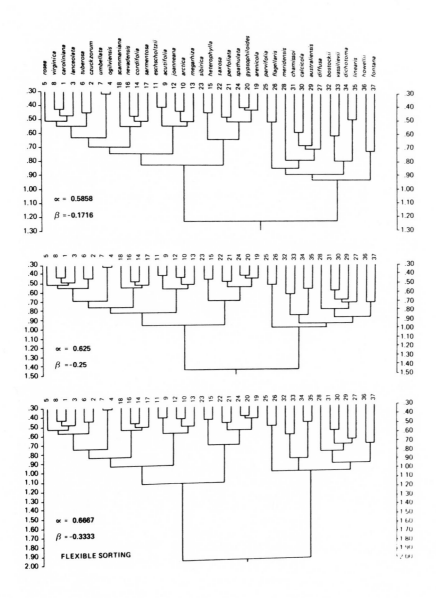

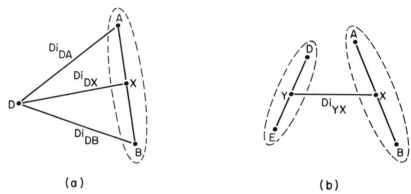

Figure 7.12 (a). Case 2. The fusion of a single item with an existing cluster using a central point algorithm such that $Di_{D,AB} = Di_{DX}$. (b). Case 3. The fusion of two clusters based on the central point algorithm that $Di_{DE,AB} = Di_{XY}$.

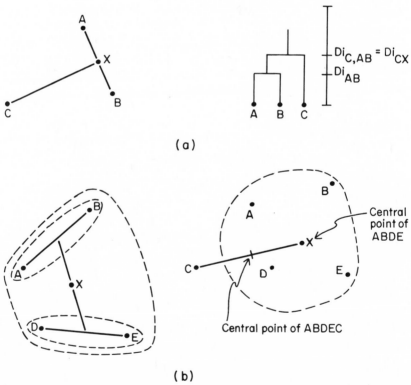

Figure 7.13 Means for finding the central point in the median cluster analysis method; see text for explanation.

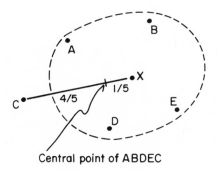

Central point of ABDEC

Figure 7.14 Means for finding the central point in the true centroid cluster analysis method.

The median cluster analysis method was described by Gower (1967b) and the true centroid method in the same year by Lance and Williams (1967). Both cause the raising of crossbars at point/cluster and cluster/cluster to even higher levels than the group-average method. Both have another property; reversals can occur; after a cluster is formed the central point of the cluster may be closer to a third point than the two cluster members are to each other. The result is a crossbar lower, not higher, than previous crossbars, as shown in fig. 7.15.

The early rejection of single-linkage by Watson, Williams, and Lance (1966 and 1967) was followed by a preference for the true centroid method because of its tendency to give apparently clearcut groupings more frequently because of what we now call space dilation. The results of the 1967 study of the order Ericales by the above authors is shown in fig. 7.16. They reproduce the traditional divisions into Ericaceae and the Epacridaceae

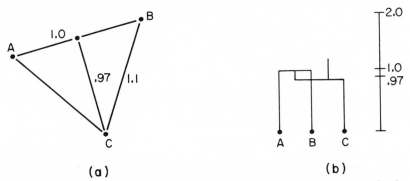

(a)   (b)

Figure 7.15 Diagrams showing means by which reversals can occur in central point clustering methods.

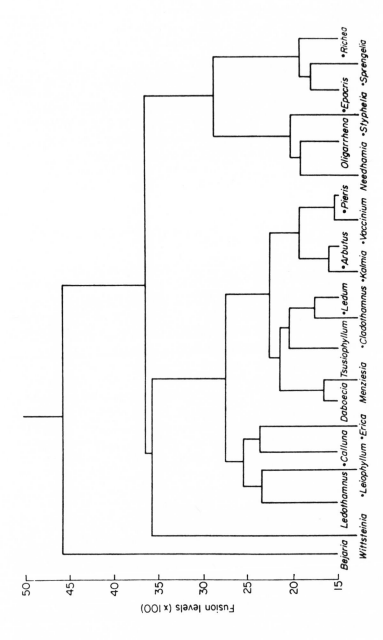

Figure 7.16 Groupings within the order Ericales produced by true centroid cluster analysis by Watson and coworkers. *(From Watson et al. 1967. Fig. 1).*

with *Wittsteinia* as an outlier, but at the lower levels produce novel groupings. In the same paper Watson et al. (1967) provide some further evidence, not included in the computer study, to support some of these novel groupings.

An interesting lesson, however, can be drawn from the severe criticism of the Watson group's novel groupings by Burtt, Hedge, and Stevens (1970). They challenge the usefulness of the results largely because of supposedly sparse and erratic sampling of the genera and of unsatisfactory character state definition. The point is now well understood—no amount of careful pattern analysis will reveal useful results if sampling and detail of observation are not of the standard expected by orthodox taxonomists.

Optimization Cluster Analysis

The reader should be aware of the existence of another family of cluster analysis methods based on optimizing the homogeneity within clusters and the heterogeneity between clusters. Despite wider use in ecology than taxonomy, on theoretical grounds agglomerative versions of optimization should be as attractive as, say, the central point methods. Several methods exist that depend on either SS (sums of squares of distances of cluster members from a cluster mean) or I (information or entropy) as indicators of the heterogeneity of a cluster. For details of these methods see chapter 5 and also Clifford and Stephenson (1975).

In the **sums of squares agglomerative method** all possible pairwise fusions are examined and the optimum fusion is chosen in such a way that making the new cluster gives the minimum rise in heterogeneity. Thus, if Y and Z are candidates for fusion into cluster X, ΔSS, the change in the sums of squares, representing the rise in heterogeneity, is minimized as follows:

$$\Delta SS = SS_X - (SS_Y + SS_Z)$$

In the **monothetic divisive method** the items are treated as one cluster, all possible monothetic divisions into two subclusters are examined and the division that gives the largest drop in total heterogeneity is chosen. A monothetic division is one based on a single character so that in the most common case of binary data the two subclusters have uniformly different scoring for the chosen characters. The drop in heterogeneity that is maximized can be ΔSS as above, or ΔI. In the latter case, where cluster X is divided into subclusters Y and Z, the formula is

$$\Delta I = I_X - (I_Y + I_Z)$$

Monothetic divisive methods have two advantages which have particularly attracted ecologists. For very large data sets the top few divisions may provide the classification wanted and the computer analysis can then be stopped before excessive time is wasted; when there are, say, thousands of items, nearly all other methods take amounts of time that can be prohibitive. Second, the monothetic classification results in a hierarchy of clusters distinguished by diagnostic character states. Taxonomists realize that monothetic divisive classifications are effectively artificial classifications, which are known to have problems. At least in theory, however, they may be useful in keymaking (see chapter 9).

Ordination

Ordination techniques are used by taxonomists to produce scatter diagrams, in two dimensions, showing items in positions which best summarize their positions in the multidimensional geometric space. Two techniques of particular utility in taxonomy that will be described here are Principal Coordinates Analysis (PCO) and Multidimensional Scaling (MDS). Also, as discussed in chapter 5, taxonomists make use of the closely related Principal Components Analysis (PCA) and, occasionally, of Factor Analysis as well.

Both PCO and MDS involve conceptually flattening the points in the multidimensional space so that they lie on a plane surface. Their relative positions on this surface are therefore related to their positions in the original multidimensional space. In the simplest case, of reducing three dimensions to two, PCO can be thought of as projecting points onto a plane. The plane surface chosen is the one which gives the largest dispersion of points so as to account for as much as possible of the variance. One can imagine it as shining a light from a distance through points arranged in three-dimensional space onto a sheet of paper and then rotating the light and paper until the best angle is found for maximizing the spread of points. In MDS the points may be thought of as marbles joined by rubber rods, each rod of length equal to the distance between two items. The whole structure is now squeezed and bent until it lies flat. The procedure, although allowing for some rods to be stretched more than others, aims at maintaining the rank order of magnitudes among the rod lengths while minimizing the total amount of stretching and squeezing.

Principal Coordinates Analysis

Principal Coordinates Analysis (PCO) is related to Principal Components Analysis (PCA). PCO allows extension of the PCA methods based on utilizing quantitative data to a method applicable to qualitative or mixed data. With PCO one can use all kinds of data provided only that measures of distance or similarity can be calculated among the items. Therefore PCO is usually used in preference to PCA in taxonomic pattern analysis.

PCO, like PCA, consists of repeated steps to locate the principal axes in multidimensional space, starting with the first and usually stopping with the second or third. The idea of locating objects with respect to successive principal axes or components has been described in the Geometric Model section in chapter 5. As the reader will recall, it consists of finding what we can loosely call lines of best fit through the points. Computations in PCO can be thought of in terms of projections of points or planes rather than as rotations of axes as in PCA. The first axis, for instance, is the one along which the projection of points has the maximum spread as measured by its having the largest share of the variance. If this line fits perfectly, passing through all of the points, the dispersion along the line (that is, the variance or sums of the squared deviations from the mean) would have a value equal to 100 percent of the original dispersion of points in the space, and the position of the points along the line would give a perfect representation of the original. Usually, however, the points are not found exactly on a line and dispersion along the axis is less than 100 percent. Clearly the higher the percentage of the original dispersion accounted for by this first principal axis, the better the approximation.

The two-dimensional scatter diagram that is usually the result of a PCO analysis is drawn through the first two principal axes, which are at right angles to each other. Points representing the items are plotted by utilizing the values computed for their projections on these axes. The simplification of the multidimensional model to a two-dimensional scatter diagram or a three-dimensional physical model means that the user can examine the collective pattern in the data visually—something that was not possible with the original data table. In general there are two kinds of visual analysis: looking for trends and looking for clusters.

If the PCO scatter diagram shows a cloud of points with no major discontinuities, then taxonomists may identify the items toward the extremities of the cloud of points. They either note that they form opposite extremes of the variation range (trends) or try to relate these extremes to extremes in other extrinsic data not included in the original data of the study.

Alternatively, the taxonomist may be able to discern clusters in

168COMPUTER-ASSISTED ANALYSIS

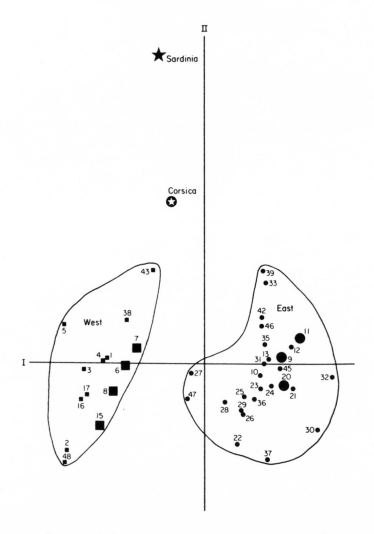

the scatter diagram. As in cluster analysis, these visually located clusters are putative classes in a classification. However attractive this method may seem, it remains very subjective. First, there is the difficulty of being certain about cluster boundaries. They can be affected by the user's subjectivity. There are variations in which way is up when the scatter diagram is drawn, whether the points are identified beforehand, and how the identification is drawn in. A second problem is simply that items close together in the scatter diagram plotted for axes I and II may be separated into different clusters when axis I is plotted against axis III. These problems are illustrated in chapter 8.

This emphasizes one of the basic problems of using PCO in tax-

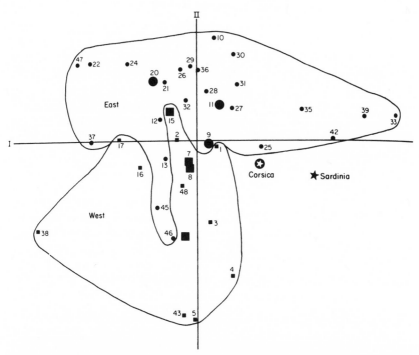

Figure 7.17 (pp. 168–169) Results of PCO used to analyze morphological diversity in populations of grass snakes. *(From Thorpe 1980. Figs. 4 and 12 respectively).*

onomic work. Because the original characters are to some unknown extent correlated, distances in the PCO scatter diagram are not directly proportional to the original distances. As a result, the method is almost a bithetic one—dispersion along the chosen axes is used to the exclusion of that along remaining dimensions. Characters whose variation does not contribute to these axes are ignored.

Thorpe (1980) used PCO to examine the morphological pattern among grass snake *(Natrix natrix)* populations from North Africa, Europe, and Asia. Fig. 7.17 (p. 168), drawn from a PCA using 56 characters from males, shows clear separation between two major clusters and indicates two outlying forms. The latter are also geographically isolated, one being from Corsica and the other from Sardinia. The sharp distinction between the two clusters is of particular interest when one realizes that the two represent an eastern and a western form, and that even though each has a range of hundreds of miles, the transition zone between them occupies a belt about 50 miles wide southeast of Lake Constance near the German-Swiss border. When Thorpe performed a PCO using only 13 scale characters, the two forms appeared interdigitated in the two-dimensional scatter diagram of principal axes

I and II, shown in fig. 7.17 (p. 169), although there is some further separation in axis III that is not shown in the illustration.

This pair of scatter diagrams illustrate, then, both the success and the problem with PCO: important gaps or distances may or may not be in the projection of the first two principal axes, particularly when the contributions of these axes are relatively low. Only in the case of separations to which all characters contribute (and hence a rather simple taxonomic pattern) is it certain that PCO will detect them.

Multidimensional Scaling

The aims of MDS (or nonmetric multidimensional scaling, its full name) are to produce new coordinates for the P items in a reduced space of t dimensions, where $t<n$, and to yield a simplified geometric model; in the case where $t=2$, it is a simple scatter diagram. MDS differs from PCO in aiming to preserve the rank order of magnitude of distances between the points, so that, for example, larger distances in the original model are larger in the simplification and so on. It is thus more truly phenetic than PCO.

There is no direct method of calculating the best answer and in a sense MDS is an exception to the claim that taximetric methods are absolutely precise and repeatable. The computer program which performs the MDS starts from an arbitrary but plausible arrangement of the points and makes adjustments to the positions until a best set of positions is encountered (Kruskal 1964). Because of an uncertainty in the technique as to whether the 'best' solution is the absolute best, the calculation is usually repeated several times, beginning from different starting positions. The best set of positions is defined as that with minimum departure from the original rank order of distances as measured by Kruskal's measure of stress, S, defined in chapter 5.

Prentice (1979) uses MDS in studying the pattern of variation within and between two species of campion *(Silene)* (fig. 7.18) and in so doing shows very clearly the two taxonomic uses of ordination techniques. For each analysis she ran the computer program twice, but found only minor differences in results. Within *S. alba* the cultivated populations scored for 34 seed, flower, and capsule characters showed a continuum with a clear geographic trend for populations from Spain and Portugal at one extreme and those from Yugoslavia and Romania at the other, as seen in fig. 7.18(a). When data on the same characters were combined for *S. alba* and *S. dioica*, they gave very clear, separate clusters, that for *S. dioica* being dense and implying homogeneity, and that for *S. alba* being more heterogeneous because of the trend mentioned above (fig. 7.18(b). This shows that on these characters the two species are distinct. The diagram in fig. 7.18(c), which is based on seed characters only, shows no clear clusters. Indeed, seed characteristics of *S. alba* are intermixed with *S. dioica,* and one must conclude that on these characters the species overlap and cannot be perfectly distinguished.

Canonical Variates Analysis (MANOVA)

Canonical variates analysis, also known as multiple discriminant function analysis, or Multivariate Analysis of Variance or MANOVA (Rempe and Weber 1972) differs from all the techniques discussed so far in that it is used to study the pattern of variation not between individual units but between samples or preclassified groups. In one sense, it can be used to test this preclassification in that it determines the level of certainty at which the preclassified groups can or cannot be distinguished. Secondly it provides an ordination of the groups or samples, each sample being represented as a cluster of points on a scatter diagram. Unlike PCA, PCO, or MDS methods, MANOVA does search for gaps between clusters.

The method depends on two effects which can be explained using the geometric model introduced in chapter 5. First, it uses the statistical Discriminant Function technique, discussed in chapter 5, to locate the best axis on which to discriminate between two clusters of points. In fig. 7.19 the best discrimination along a single axis is obtained by projecting all the points onto line *A,* which runs through the means of both clusters. But in MANOVA the problem is more complicated in that line *A* must maximize the dispersion between not two clusters but many.

Second, the technique takes into account not only the magnitude of variation in each cluster (as above) but also any correlations between character variations within the clusters. Such correlations cause clusters to be attenuated in the same direction in the geometric model. If the attenuation is, for instance, perpendicular to the selected line as in fig. 7.20(a), then it has no effect on the discrimination. However if, as in fig. 7.20(b), the attenuation is in the direction of the selected line, then the discrimination is reduced. The MANOVA takes this covariance into account by first transforming the data (in this case the model) to make as many as possible of the clusters spherical. In fig. 7.20(a) the squeezing will be perpendicular to line *A,* thus not affecting the canonical distance between clusters, but in fig. 7.20(b) the squeezing will be along line *A,* so reducing the canonical distance. The canonical distances between clusters obtained in this way are known as Mahalanobis or D^2 distances. (See chapter 5)

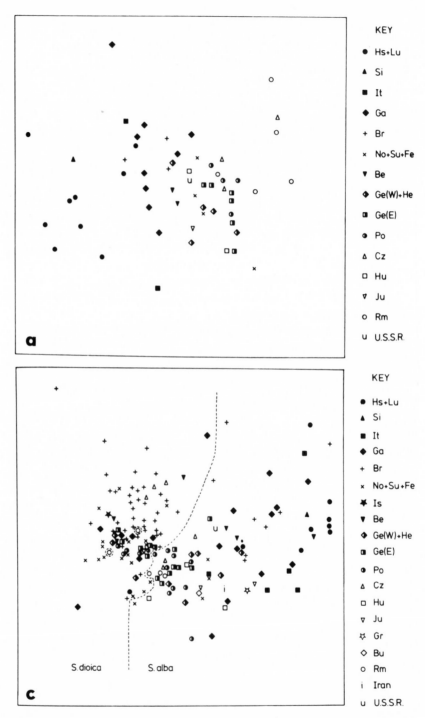

Figure 7.18 Results of MDS used to analyze variation between two species of *Silene;* see text for explanation. *(From Prentice 1979. a, Fig. 2; b, Fig. 8; c, Fig. 9; Table = Table 2, p. 191).*

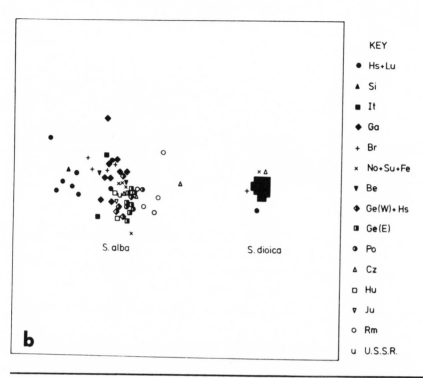

KEY

● Hs+Lu
▲ Si
■ It
◆ Ga
+ Br
× No+Su+Fe
▼ Be
◈ Ge(W)+Hs
▣ Ge(E)
◐ Po
△ Cz
□ Hu
▽ Ju
○ Rm
ᴜ U.S.S.R.

S. alba   S. dioica

b

Abbreviation	Territory
Be	Belgium and Luxembourg
Br	Britain (England, Wales and Scotland, including Orkney and Shetland Islands)
Bu	Bulgaria
Cz	Czechoslovakia
Fe	Finland
Ga	France
Ge(E)	E. Germany (German Democratic Republic)
Ge(W)	W. Germany (Federal Republic of Germany)
Gr	Greece
He	Switzerland
Hs	Spain
Hu	Hungary
Is	Iceland
It	Italy, excluding Sicilia
Ju	Jugoslavia
Lu	Portugal
No	Norway
Po	Poland
Rm	Romania
Si	Sicilia
Su	Sweden, including Öland

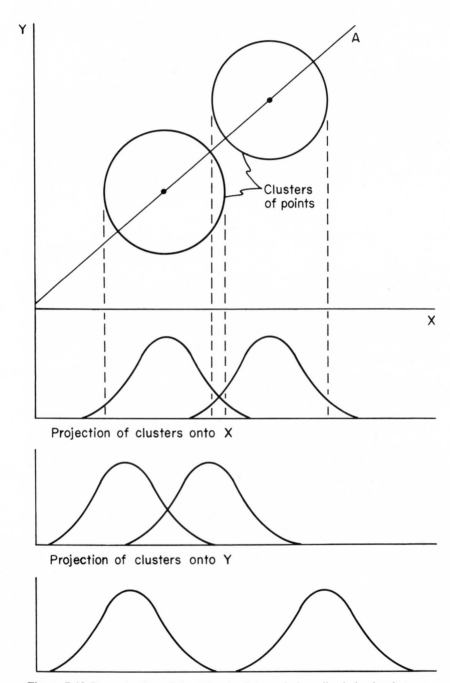

Figure 7.19 Determination of the axis, A, that maximizes discrimination between two clusters using Discriminant Function technique.

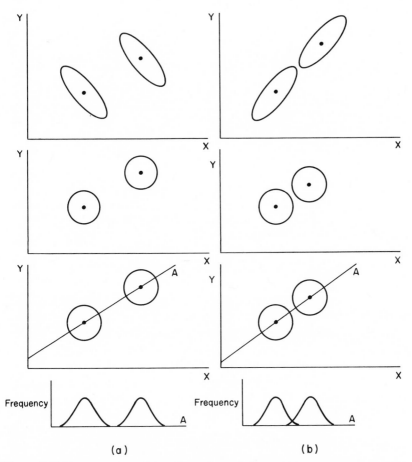

Figure 7.20 Demonstration of the effect of correlations within clusters on canonical distances between clusters in MANOVA.

        The results of canonical variates analysis are drawn on a scatter diagram. One method is to draw the clusters as circles with a mean placed on a scatter diagram. If various assumptions are met (that each cluster is multinormal and that the variance and covariances are similar in all samples), one can draw the circles at 95 percent confidence limits around the means; that is to say, 95 percent of units sampled for one cluster would fall within the circle. Alternatively, one can draw the original data units on the scatter diagram. An empirical test will then show what percentage of the original units would be correctly assigned if they were assigned to the cluster with the nearest mean. The method is thus a statistical one expressly accommodating variation or sampling error in the original samples.

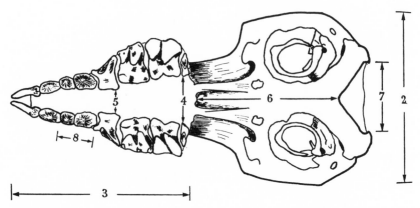

Figure 7.21 (a, above) Drawing of skull of a shrew indicating measurements used in the canonical variates analysis. (b, upper right) Map showing distribution of the shrew populations. (c, lower right) The results of canonical variates analysis showing groupings of the various island populations on the basis of the skull measurements. *(From Delaney and Healy 1966. a, Fig. 2; b, Fig. 1; c, Fig. 5)*

Canonical variates analysis has been put to particularly effective use in studying the variation at and below the species level in populations of organisms in which the variation occurs in numeric characters only. Fig. 7.21 illustrates the pattern, based on certain skull measurements, in samples of white-toothed shrews on the islands in the British Channel (Delaney and Healy 1966). The accuracy of the circles enclosing 90 percent of the animals in each sample depends on the variability being similar in each population. The result illustrates that certain populations can or cannot be distinguished: for example, the overlap of the Guernsey *(G)* circle with the Cap Gris Nez *(C)* circle means that 90 percent of the Guernsey animals are not distinct from 90 percent of the Cap Gris Nez animals, but the separation from the Jersey *(J)* sample means that these can be clearly distinguished from Jersey animals. In addition to providing a statistical test of distinguishability, the arrangement of the sample means acts as a summary of the variation pattern between samples. Thus, while the Jersey and Sark *(S)* populations are statistically indistinguishable from the Scilly Islands' populations [Tresco *(T)*, St. Mary's *(My)*, St. Martin's *(Mn)*, St. Agnes *(Ag)* and Bryhyer (B)], they are more similar to each other than they are to the Scilly populations. This ordination differs from that produced by PCA and PCO in that it is an ordination of the sample means after the effects of variability within samples have been taken into account, and with the express intention of maximizing the separation of samples. The authors use this analysis to describe the variation pattern in the two species involved, *Crocidura russula* on Alderney *(Al)*, Guernsey and Cap Gris Nez (mainland France), and *Crocidura suaveolens* on the remaining islands.

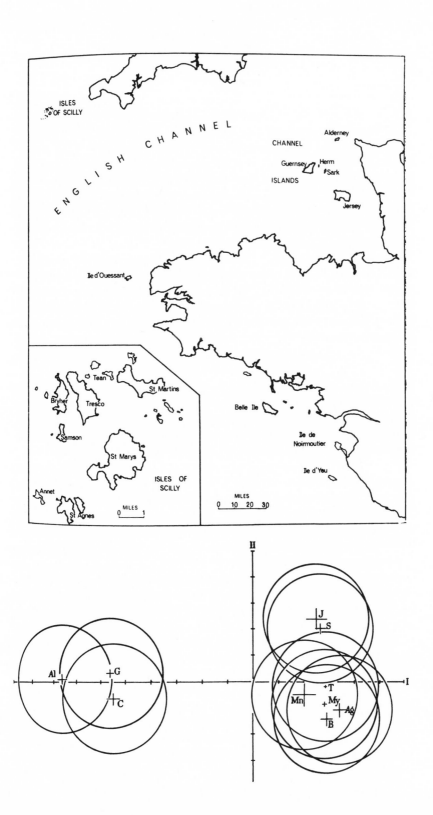

Choice of Methods

It is important to remember that at no time before or after the arrival of computer-assisted taxonomy has there been a single, precise method by which orthodox taxonomists have agreed to analyze phenetic patterns and classify from them. It is little wonder, then, that the wide range of loosely defined procedures in use has led to an enormous range of precise computer methods from which the modern taxonomist must choose. Those described above are in fact just a selection from those available. Two means of choosing are possible. One can choose on theoretical grounds from the logical properties of the methods or on empirical grounds from convenience and the qualities of the results. In both cases the choice may be either conservative, trusting that tried methods or established results are worth emulating, or radical, aiming for improved logic and improved results.

Logical Properties

We list here a number of important properties which vary among the methods already described. For more detailed, sometimes mathematical, comparisons see Jardine and Sibson (1968) and Williams, Clifford and Lance (1971).

Chains and Dense Clusters

In geometric terms the clusters detected by single-linkage may have any shape provided that they are not dissected by a gap. So the clusters may be hyperspherical, or long straggly chains, or dumbbells (two or more dense regions joined by one or more bridging chains), or kites (dense regions with chains attached to them like streamers). These various shapes are illustrated in fig. 7.22. In the two-dimensional example shown only the dense cluster would be detected as a cluster with the same membership by the central point method. The chain and dumbbell would be split into two or more

Dense Cluster   Chain   Dumb-bell   Kite
Figure 7.22 Various cluster shapes.

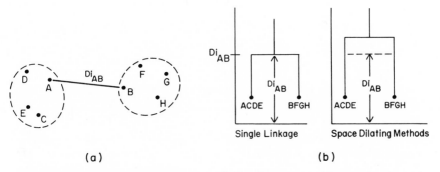

(a)   (b)

Figure 7.23 Example illustrating space dilation that occurs with central point and group-average clustering methods.

clusters and the outliers on the streamer of the kite might well have been omitted. An argument often raised against single-linkage is that from a dendrogram one can not tell which clusters contain chains (or necks, waists, or promontories). However, such cluster shapes are easy to detect if linkage diagrams are drawn.

Space Dilation and Space Contraction

When two clusters are separated by a gap Di_{AB} between item **A** in one cluster and item **B** in the other, they fuse at level Di_{AB} in single-linkage. In group-average and the central point methods they fuse at a level greater than Di_{AB}. This effect, which makes clusters seem more distant, is called **space dilation** and is illustrated in fig. 7.23. It is most extreme for the median method, less extreme for the true centroid method, and least for group-average cluster analysis.

In a single-linkage cluster analysis two members as **F** and **G** in fig. 7.24 at opposite extremes of a chain, join the same cluster at a distance less than Di_{FG}. The intermediate chains make them seem closer: this is called **space contraction.**

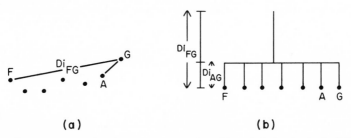

(a)   (b)

Figure 7.24 Example illustrating the space contraction effect of chains in single-linkage clustering analysis.

Gaps and Partitions

In single-linkage, clusters are delimited by the gaps around them. The size of the gap can be read off from a dendrogram or determined from a series of linkage diagrams. Thus, in fig. 7.25 clusters **AB** and **CD** are both isolated by gap α. In the space dilating methods α is not the size of a gap between **AB** and **CD**, indeed the gap may be less than Di_{DC}. It is in this sense that space dilating methods introduce exaggerated gaps between clusters.

When choosing between PCO and MDS it should be borne in mind that the Principal Axes of PCO are chosen to maximize dispersion, *not* to maximize gaps between clusters of points. Indeed large gaps, if they occur in characters otherwise of low variance, may not be incorporated in the first few axes. This problem should be less prominent in MDS, where inter-point distances tend to be maintained in the rank order of magnitude.

Chaining and Balanced Hierarchies

The dendrogram in fig. 7.26(a) shows **chaining**—that is, at each of the several levels a single cluster has a single item added to it. By contrast the dendrogram in (b) shows less chaining and has a more balanced hierarchy. In (a) any level above Di_{AB} reveals one cluster plus some unclustered items, whereas in (b) levels immediately above Di_{AB} show two clusters.

In the analysis of data sets which lack major discontinuities—that is, gaps in the geometric model—single-linkage may give a chained dendrogram where the more space dilating methods dilate minor gaps to give a balanced dendrogram.

Nearest Neighbors

Of the cluster analysis methods described, only single-linkage always places an item in the same cluster as its nearest neighbor in the geometric model. Such a requirement is of course part of the single-link axiom.

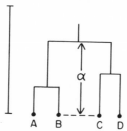

Figure 7.25 Designation of the α gap between clusters in a dendrogram.

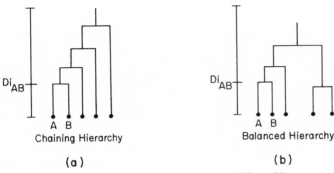

Chaining Hierarchy   Balanced Hierarchy

(a)   (b)

Figure 7.26 Illustration of chaining versus balanced hierarchies.

In space dilating methods, space dilation may occur between a given item and its nearest neighbor so leading them into separate clusters.

Stability
       In general both the addition of new items and changing the data for items will produce changes in the results. Single-linkage and the median method give unchanged results if an added item is identical to one of those already present. Single-linkage is particularly sensitive to the introduction of "missing link" intermediates between clusters, which cause the clusters in question to fuse at a lower level. The space dilating methods have been criticized for being unstable for minute changes in the distances in some configurations. Consider the distance table in fig. 7.27 where ϵ is a small change in distance. As it goes from being slightly positive to slightly negative, the single link dendrogram changes continuously, but, for instance, the group average method dendrogram changes abruptly at $\epsilon = 0$ with the so-called see-saw effect—a jump of two steps of .5 in the heights of the crossbars shown in fig. 7.27.

Cophenetic Correlations
       Sokal and Rohlf (1962) introduced the use of the product-moment correlation coefficient to measure **cophenetic correlations**—correlations between the distances implied in cluster analysis or ordination results and the original distances in the distance matrix on which the analysis is based. In a dendrogram the distance implied between two items is the height of the dendrogram node joining the two items. In an ordination the distance implied is that between points on the resulting scatter diagram or, in cases where more than two axes are used, the resulting space. A triangular table of implied distances between items is constructed and then values in this table and corresponding values in the original distance table are paired and a

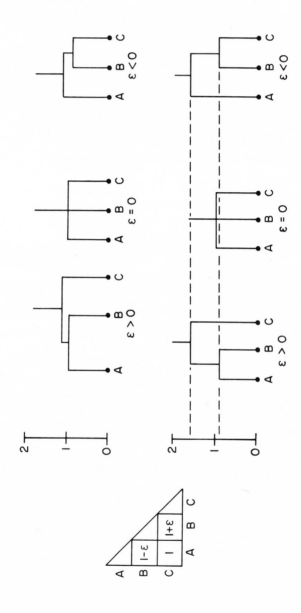

Figure 7.27 Illustration of the effect of small changes (ε) in distances on single-linkage (top) and group-average (bottom) dendrograms.

product moment correlation coefficient, r_c, calculated. Farris (1969) showed mathematically what has been known for some time from empirical studies—group-average cluster analysis tends to give high values of r_c.

Whether or not r_c gives taxonomists a useful measure of how faithful a result is to the data is a matter for debate (Gower 1983). Not only does it assume that all distance values should contribute to the evaluation but also (because of properties of correlation measures) it is very sensitive to the faithfulness of a result for the very high and very low distance values.

A possibly extreme taxonomist's view might be that discontinuities (i.e., gaps in the model) alone should be used to delimit clusters and that the breadth of variation in a taxon (distances between cluster members) might vary. The holder of such a view would presumably approve of a method which faithfully reproduced the small distances to get the gaps and nearest neighbors accurately mapped, but might be unconcerned about the treatment of many medium and most of the large distances.

S, the stress measured in MDS ordinations, and SS, the total dispersion accounted for by the principal axes of a PCO, are both other measures of cophenenetic faithfulness.

Comparing Classifications

There is a considerable literature on quantitative methods for comparing hierarchical classification (Rohlf 1974). The results of all of the methods are difficult for taxonomists to evaluate, and only one has been widely used—a variant of cophenetic correlation. It is applied between the ultrametric distances of two dendrograms for the same items either produced by different methods for the same data, or between dendrograms for the same items from different data using the same method. A more recent development is the idea of building consensus trees (fig. 7.28) showing the elements of agreement between two trees, or of computing a consensus index between the two trees under comparison (Adams 1972; Mickevich 1978; Rohlf 1982; Gower 1983).

Group-Size Dependence

Williams, Clifford, and Lance (1971) discuss the choice between so-called weighted and unweighted cluster analysis methods in terms of group-size dependence. When deciding on the fusion of two clusters, or the position of the central point once they have fused, does it or does it not matter that one is, say, a singleton and the other is a cluster with many members. In taxonomic terms this might be two putative orders, say the Coniferales with hundreds of species and the Ginkgoales with its single species. Alternatively, it might be two British species of *Genista:* one, *Genista tinctoria,* represented by many samples because of its widespread distribu-

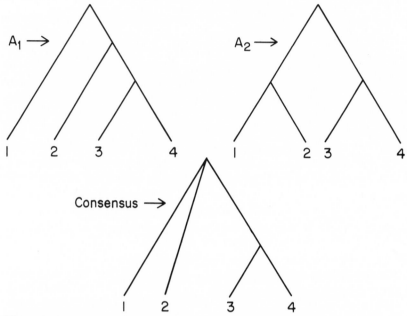

Figure 7.28 Example of consensus trees. *(After Gower 1983:146)*

tion or variable form and one, *Genista pilosa,* represented by a single sample either because of its rarity or its lack of local variation. The question revolves around whether one treats the two taxa as two equivalent entities, or as one large and one small group of members.

Where the group-size variation may be the consequence of erratic or uncontrolled sampling, it clearly should not be taken into account. This would be an argument, for instance, in favor of the median cluster method. But when disparate group sizes are a genuine feature of nature, the question is more difficult to answer. In theory at least, taxonomists catalogue the various kinds of organisms regardless of how well represented they are and so again the answer would be to choose a method that gives joining clusters equal weight; but yet, in contrast, the very notion of "homogeneous clusters of similar forms" is dependent on the presence of many members. Reduce the members to two and "cluster" may seem neither homogeneous nor even distinct! The notion of taxonomic correlations of characters also assumes a contribution from replicates of certain character combinations. The question underlines a further difficult theoretical and practical problem. What is it that taxonomists must sample—the world's surface (area), the world's taxa (species) or the world's plants (individuals)? If the answer is 'taxa,' then how can this be done before determining what are the taxa?

Empirical Properties

Single-linkage was one of the first cluster analysis methods to be tried, and in many studies one of the first to be rejected on the grounds that chaining often leads to the separation of insufficiently distinct clusters (Watson, Williams, and Lance 1967). However authors who have used linkage diagrams for examining single-linkage results seem to have been happier with these results than with other methods (Prance, Rogers, and White 1969; Stearn 1971; Bisby 1973). But using linkage diagrams does require more detailed work with the results of the computer analysis.

Many authors have shown preferences for the results after using the true centroid and median methods. They have been attracted by the balanced dendrograms and described the methods that better separate distinct clusters as more powerful. Disadvantages have been the occasional discovery of dump clusters (late-forming clusters made up wholly of outliers not closely similar to each other) or of exaggerated cluster differences. Both are, of course, the consequence of space dilation.

For first-hand accounts of comparative studies using cluster methods above the species level, see Prance et al. (1969), Bisby (1973), McNeill (1975) (all Angiosperm studies); Sokal and Michener (1967), Boratynski and Davies (1971) (both insect studies); and Boyce (1964 and 1969) (Hominoid studies).

Ordination techniques have been more widely used in subspecific studies, where their ability to depict continua and their partial ability to detect clusters is of particular value. The methods preferred and used have tended to reflect the historical sequence in which the methods became available rather than a changing consensus of preference. Until the mid 1960s, variants of PCA were used, although some argued that the methods were suitable for quantitative and ordered multistate data only. Then in 1966 Gower introduced PCO, in which a distance table calculated from any type of data could be analyzed, and this became widely used. MDS was introduced by Kruskal in 1964, but received little attention from taxonomists until the 1970s, and there are relatively few comparative studies such as those of Rohlf (1972) and of Thorpe (1980). These authors preferred MDS and PCO respectively, so leaving the matter open. Indeed it may be that, as with cluster methods, each has its own advantages or that its advantages are brought into use in different taxonomic situations. An additional study of interest for the range of techniques used but not compared is that by Jardine and Edmonds (1974).

Recommendations

Above the Species Level

Our principal recommendation is to use single-linkage with linkage diagrams above the species level. The data will normally include a high fraction of qualitative characters and we recommend the appropriate mixed-data coefficient. Linkage diagrams will reveal the full diversity of clusters, chains, and outliers. Only when the pattern revealed is effectively unclassifiable, as when all items fuse or chain onto one cluster, might it be useful to try a space dilating method such as true centroid cluster analysis. Then one must be aware that the pattern is being exaggerated to produce artificially classified groupings and that there are attendant problems. An alternative method of classifying the result is to make a stricter selection of characters with high relative information contributions.

The group-average method should be recommended only as a quick and easy look at a variation pattern where the user is not interested in the fine detail of cluster shape and delimitation. It avoids the labor of preparing and interpreting linkage diagrams, but at the expense of precision and lost information.

Below the Species Level

Below the species level, the taxonomist tempers the search for a nonoverlapping classification with knowledge that this ideal is frequently not possible. First, we are often limited to continuous characters, and indeed these continuous characters may show high levels of variability within populations. Second, we expect variation within species to involve either a continuum or at least some continuity or overlapping between forms. The first of these must be overcome by a carefully coordinated choice of characters, sampling of individuals, and choice of resemblance measure; the second by the use of an ordination technique or a cluster analysis technique that is sensitive to continua. We recommend the use of canonical variates analysis, where the data are continuous measurements and where the items are preclassified as in samples, biological populations, or postulated groupings. Where this is not feasible, we tentatively suggest MDS. Combined tactics involving variants of single-linkage may be of use here; that is, combining the ordination and classification functions. Linkage diagrams, minimum spanning trees, or Jardine's B_2 nonhierarchic clustering (Jardine and Sibson 1971) can all be superimposed on MDS or PCO scatter diagrams.

Suggested Readings

General

Dunn, G. and B.S. Everitt. 1982. *An Introduction to Mathematical Taxonomy*. Cambridge: Cambridge University Press.
Everitt B.S. 1974. *Cluster Analysis*. London: Heinemann.
Gordon, A.D. 1981. *Classification*. London: Chapman and Hall.
Sneath, P.H.A. and R.R. Sokal 1973. *Numerical Taxonomy*. San Francisco: W.H. Freeman.

Discussion

Duncan T. and R.R. Baum 1981. Numerical phenetics: its uses in botanical systematics. *Ann. Rev. Ecol. Syst.* 12:387–404.
McNeill J. 1984. Taximetrics Today. In V.H. Heywood and D.M. Moore eds. *Current Concepts in Plant Taxonomy*, pp. 281–299. London and New York: Academic Press.

Chapter 8

Diagrams
of Variation Pattern

Diagrammatic representation has played an important part in communicating taxonomic results for several centuries. Diderot (1751 *et seq.*) included a leafy tree diagram in the frontispiece of his *Encyclopedie* depicting the classification of all knowledge. All of the explicit techniques for analyzing phenetic patterns described here produce results which are commonly represented by diagrams, whether drawn by hand or computer. These diagrams are used for communication, and consequently their nature and interpretation are important matters as much for those who use taxonomy as for taxonomists themselves.

Some diagrammatic styles are tied to particular classification or ordination methods. The aim is to provide the clearest, simplest diagrams that will convey as much of the phenetic pattern information as is needed. Usually a compromise must be made between the amount of information displayed and the need to avoid complications which might detract from the principal features and complicate the interpretation. Users may well be at different levels of sophistication and thereby require different diagrams. The importance of diagrammatic display has probably been underestimated, and research into new styles is needed. It is equally important to remember that diagrams are abstractions, as are all models, and therefore only partial representations of reality.

Scatter Diagrams

A simple way of displaying the pattern found in two quantitative or ordered multistate characters is to draw a two-dimensional scatter diagram. The diagram may reveal some elements of pattern which were not revealed by examining the two characters independently. For instance, in fig. 8.1 from Agnew (1968), the discontinuity between the rush *Juncus conglomeratus* and plants intermediate with *J. effusus* is not evident in either character taken singly.

Two-dimensional scatter diagrams are drawn by placing points representing items on a coordinate diagram with the two characters as axes. The points are placed using data values for the two characters as x and y coordinates. As the pattern observed is affected by arbitrary choice of scale, a common practice is to standardize values for each character beforehand (see chapter 4).

Patterns on the scatter diagrams are recognized subjectively by the user. In many studies where the investigator has prior knowledge about the items it is important to avoid bias by performing this visual recognition of clusters or patterns on scatter diagrams in which the points are not labeled. The user draws lines on the diagram delimiting clusters or trends, and only after this is complete should one identify the points with the identifying symbols. The reason for this is that ambiguous circumstances arise in which it is very difficult not to enhance the result where it might agree with a prior classification or pattern known to the user. Also, identifying symbols on the diagram can blur or alter the visual impression of the pattern.

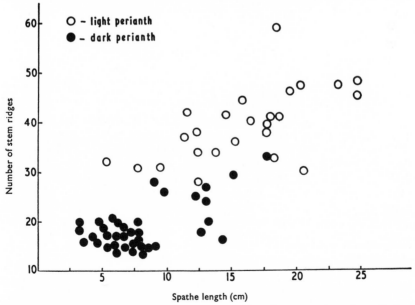

Figure 8.1 A two-dimensional scatter diagram illustrating separation of two distinct groups of plants on the basis of two characters. Filled circles are *Junceus conglomeratus;* open *J. effusus. (From Agnew 1968. Fig. 2).*

Most computer installations provide facilities for drawing 2-D scatter diagrams automatically. At their most crude, these may be produced by printing the symbols on successive lines, either on a line printer or on a video display screen. Such diagrams are easy to produce. The main problem with them is the crude positioning of points, which are rounded to the nearest line and row position for a print symbol, and the fact that this rounding increases the chances that several symbols will fall on the same position, causing further printing and labeling difficulties.

Better two-dimensional scatter diagrams can be generated by the various drawing, photographic, and video-screen graph plotters on which the pen or pointer can move to any exactly defined position. There have been enormous advances in the graphics facilities available with microcomputers and what were once expensive and time-consuming facilities on mainframes are now widely available to give a product that is generally more pleasing. Often a wider variety of symbols is available, the diagram may be more accurate, and minor adjustments in labeling and separation of overlapping symbols may be made. Fig. 8.2, from Sokal and Rohlf's (1980) studies on taxonomic judgment using manmade "caminalcules," shows how scatter diagrams produced by graph plotters with fine, accurately positioned pens suffer the slight disadvantage of being barely bold enough for reproduction or publication.

Multivariate Data Tables

In theory, 2-D scatter diagrams can be used to examine the whole of a multivariate data table. In practice they are used only to examine pairs of characters of particular interest or to examine very much simplified multivariate models, such as the results of a PCA or PCO analysis. To examine a full data table requires an enormous number of diagrams [$p(p-1)/2$ for p characters, i.e., 45 for 10 characters, 4590 for 100 characters], and deciphering the multivariate pattern by eye is impracticable for more than three or four characters.

Two particular kinds of pattern are easily detected visually: long sequences of points, often called trends, and clusters. The trends are extended areas of points which stand out visually in contrast to sparsely filled or empty adjacent areas. These are interpreted as trends in the variation of the items, interesting to the user because of the lack of discontinuities or because of the interest in what causes the trend. If the trend is linear then the two characters used are linearly correlated and the user may again be

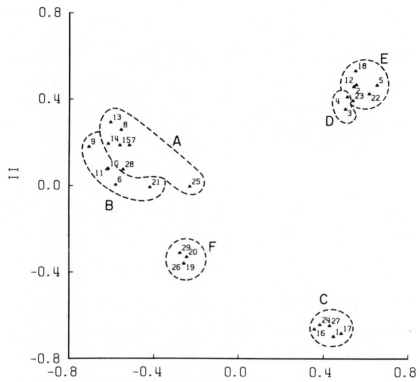

Figure 8.2 An illustration of a scatter diagram drawn by a computer graphics plotting program. *(From Sokal and Rohlf 1980. Fig. 3).*

interested in the causes. Cain (1977) has used the occurrence of two distinct trends in mollusc shell forms to hypothesize and suggest experiments on the pressures of natural selection on the shapes (fig. 8.3).

Clusters are areas densely filled with points that can be distinguished from surrounding areas, which are sparse or empty, as in fig. 8.1. Their visual recognition is subjective and attempts to systematize the recognition have led directly to the invention of cluster analysis methods and the need to choose particular criteria, such as the existence of gaps or the homogeneity of clusters.

Most PCA and PCO results are presented as one or more 2-D scatter diagrams of the kind described above. Each diagram is constructed in the same way, using coordinate values for just two of the principal axes produced in the analysis. Often several scatter diagrams are produced as for example Principal Axis I compared with II, I with III, and II with III. The differences appear in the interpretation. The first stage, again, is for the user

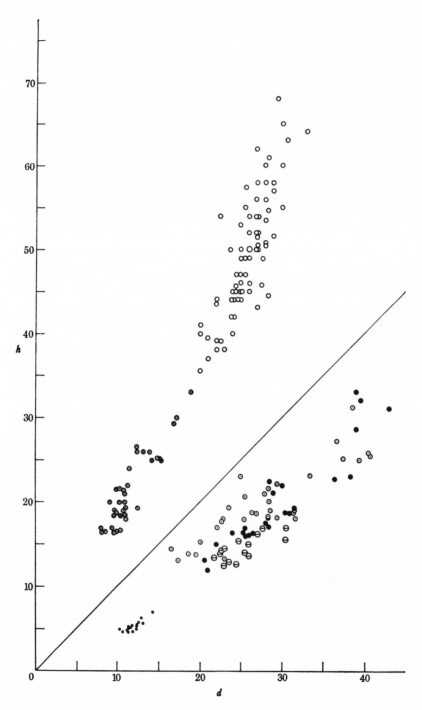

Figure 8.3 Scatter diagram showing trends in gastropod shell dimensions. *(From Cain 1977. Fig. 2).*

to detect clusters or trends by eye in each diagram and again it is important to do this in the unlabeled diagram.

Then comes the more difficult part, understanding the trends and clusters in three or more dimensions. Let us take the example mentioned above, where the first three principal axes are used to produce three 2-D diagrams. Several situations can occur:

1. If some items are in the same cluster in all three diagrams, then they are together in all three dimensions.

2. If some items are in the same cluster in two diagrams only, they are separated in one plane.

3. If some items are in the same cluster in only one diagram, they are separated in two planes.

4. If some items are isolated in all three diagrams, they are isolated in all three dimensions.

Occasionally authors give even more projections where the first three principal axes do not account for a high fraction of the dispersion. However, it is no simple matter sorting out the clusters even from just three 2-D scatter diagrams. For instance, how many clusters are there in the example illustrated in fig. 8.4?

From this example one can see another problem which limits the value of PCA or PCO results. Each 2-D scatter diagram is an approximation of, say, a 3-D simplified model. Any two items close together on it may in fact have a large distance from each other (i.e., low similarity) in another plane. If the principal axes are taken into account for a large fraction of the total dispersion, then this effect will only occasionally be of importance— as, say, in the results of Boyce (1964) where the first three axes in a PCA of hominoid skulls accounted for 66 percent. The effect is more pronounced where the approximation is poor, as in Bisby (1973) where the first five axes for a PCO of *Crotalaria* species accounted for only 41 percent. Later in this chapter we discuss the use of minimum spanning tree superimposed on scatter diagrams to reduce this effect.

Physical Models

A direct and effective way of displaying a 3-D scatter diagram is to make a physical model. A photograph in Boyce (1969) shows a simple model with labeled plastic balls supported on rods above a polystyrene base (fig. 8.5). The second and third principal axes are marked on the base, and the vertical axis is the first principal axis for PCA results for 20 hominoid skulls. Such

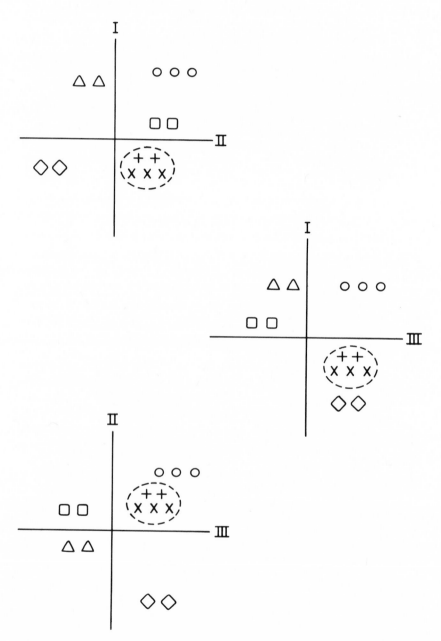

Figure 8.4 A hypothetical case that illustrates clustering based on three principal axes by plotting, in succession, their three possible pairwise combinations.

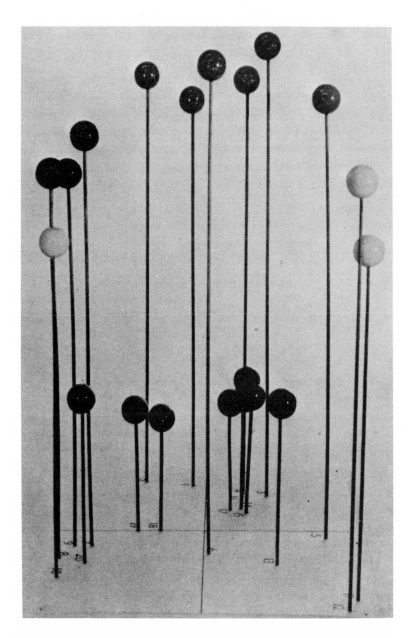

Figure 8.5 A solid model that represents positions of objects in three dimensions. *(From Boyce 1969. Fig. 9 p. 20).*

models can show quite clearly clusters and trends which are difficult to observe from a series of 2-D scatter diagrams by the methods described above.
In addition, the user can rotate the model and inspect complicated or hidden
clusters from several angles. The limitations, of course, are that the model
is difficult to make for large numbers of items and is of little use for publication or for illustrating lectures. There are also problems with positioning
balls whose centers are very close to each other.

Some of the earliest extensions of scatter diagrams to three or
more dimensions were by Anderson (1949). In his pictorialized versions,
variations in the symbols and in the flags attached to them indicate qualitative third and fourth characters. The example in fig. 8.6 shows this method
in use. More recently Gower (1967a) has attached stems to the symbols, the
length of the stem representing positions in the third dimension. The reader
might imagine these as showing how far above or below the paper the point
should be.

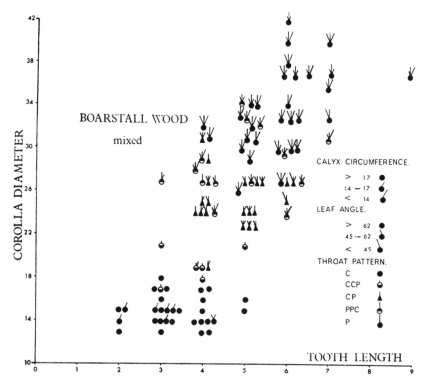

Figure 8.6 A pictorialized scatter diagram in which use of symbols extends to five
the number of characters of a hybrid population represented. *(From Woodell 1965. Fig.
3).*

Rohlf (1968) and Moss (1967) first produced perspective views of points placed in a 3-D space drawn by computer. They are normally drawn as if they were perspective views of physical models, with rods supporting discs. The discs are drawn getting smaller toward the back of the model, so that discs in front partially obscure those behind. The clarity of the 3-D impression and the extent of foreground discs obscuring others is affected by the angle from which the perspective view is created (example in fig. 8.7). In some cases the user can select what rotation is used for the perspective view.

Rohlf (1968) also produced stereographic pairs of perspective views. When viewed through a stereoscope, a stereoscopic view with an impression of distance, as well as position of the items, is obtained (fig. 8.8). Crovello (1968) has made use of another perspective model where points are projected onto the surface of a sphere which is then drawn in perspective view.

Dendrograms

A dendrogram is a diagrammatic representation of a strictly hierarchical classification. It is a tree showing progressive inclusion of items or groups of items into larger groups in a hierarchy. Items or groups are depicted by stems. The inclusion of several items or groups into one group occurs at a node, and is depicted by a crossbar as shown in fig. 8.9. At one hierarchic level or another, all items become included in groups, and these in turn all become members of the complete group. The height, measured on the ordinate, at which each node occurs is known as its rank, and is based on a measure of resemblances which varies from study to study. The different criteria for inclusion in a group in the hierarchy produce the variety of clustering algorithms discussed in chapter 7.

At each node of the dendrogram there is one larger group known as the supergroup, which is made up of two or more subgroups. To draw the node on the dendrogram, the supergroup, the subgroups, and the rank must be known. Dendrograms are easier to draw starting from the largest cluster and working through the successive nodes to the separate individuals. Divisive clustering computer programs normally produce information in the correct order and drawing is straightforward.

Many agglomerative clustering computer programs produce information in the reverse order, and if the dendrogram is drawn directly from this complications result. The problem is that the first clusters of items are drawn on the paper in an arbitrary position. Later the clusters which fuse

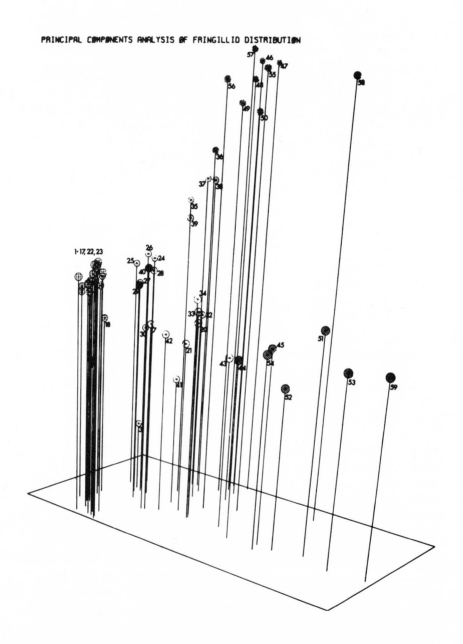

Figure 8.7 A perspective view of a three dimensional plot drawn by computer. *(From Bock, Mitton, and Lepthien 1978. Fig. 5).*

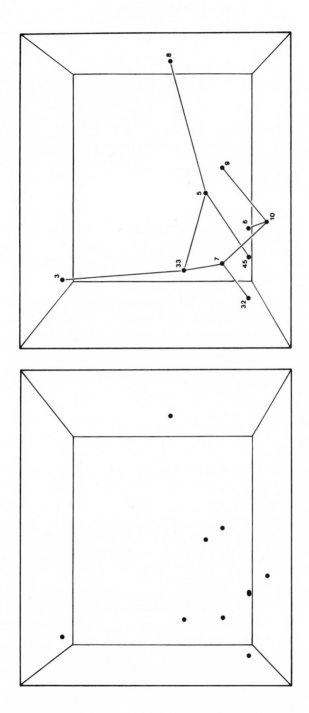

Figure 8.8 Stereoscopic pairs generated by a computer to assist in the analysis of points plotted in three dimensions. *(From Moss, Peterson and Atyeo 1977. Fig. 10).*

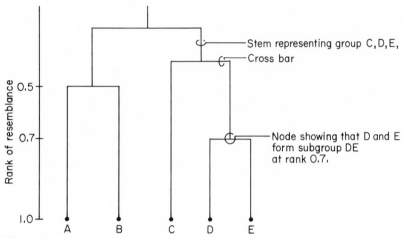

Figure 8.9 An explanatory diagram indicating the components of a dendrogram.

may not be those drawn next to each other. The result is a dendrogram with crossing stems. For a small dendrogram these crossed stems can be disentangled and a final copy produced. Alternatively the problem can be solved by storing all information about all nodes and then drawing the·dendrogram

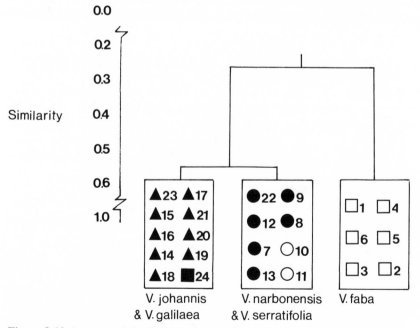

Figure 8.10 A truncated dendrogram showing the mains groupings of a set of specimens of *Vicia* section Faba. *(From Birch, Tithecott and Bisby 1985.)*

as if it were divisive, starting with the largest cluster and finishing with individual items.

An additional advantage of drawing the dendrogram in the divisive direction is that the operation can be terminated at any chosen rank or
number of groupings. Particularly in probabilistic studies there is no need to
know the exact position of every item; it is the major groups which will be
used as in fig. 8.10. In other cases the number of items may be large and
the dendrogram truncated for simplicity's sake.

In most dendrograms the ordering of items within sets is arbitrary and not intended to give meaning either within or between sets. Thus
in fig. 8.11, the order of attachment to a crossbar can be shuffled without
changed meaning. In (a) the juxtaposition of B_3 and B_4 does *not* mean that
B_4 is more similar to B_3 than to B_2; and similarly B_5 near B_4 does *not* mean
that B_5 is the most similar item of set α to B_4 in set β. This contrasts with

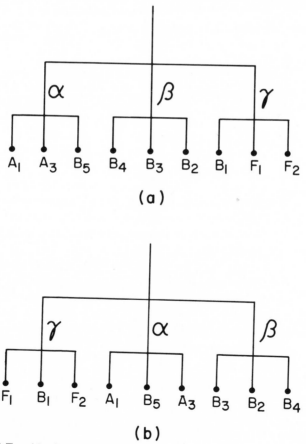

Figure 8.11 Two identical dendrograms.

the layout often used by taxonomists where the most similar taxa are printed adjacent to each other. For example, Clapham, Tutin and Warburg (1962) list the genera *Vicia* (vetches), *Lathyrus* (wild peas), and *Pisum* (true peas) in that order because although all these are members of the same tribe, *Vicia* and *Lathyrus* are very similar and of the two, *Lathyrus* is the more similar to *Pisum*.

Linkage Diagrams

Linkage diagrams, sometimes known as subgraphs (see chapter 5 for the theoretical background), have been extensively used to show more detail of variation patterns than can be shown on a dendrogram (Irwin and Rogers 1967). They are particularly useful for showing features of single linkage which may be of interest to the user and the item-by-item progress of the clustering process.

A series of linkage diagrams can be drawn, one for each rank of a single-linkage dendrogram, starting with single items and finishing with the cluster which includes all items. Each diagram is formed from the previous one by adding items. In addition to the usefulness of the final product, the very process of drawing a linkage diagram can be useful to the investigator in demonstrating the details of the interrelationships among his objects. We will detail step-by-step the production of a linkage diagram with an example for which the similarity matrix is shown in fig. 8.12.

Readers will recall that single-linkage agglomerative clustering is performed by progressively searching the similarity table (see fig. 8.12) for the next highest value. Draw the first linkage diagram when the highest value S_{IJmax} has been located. In this case it is .95 for two pairs of items (1 and 2, 4 and 5). Place these pairs of points in an arbitrary position on the diagram representing each item by its identifying number and joining the members of each pair by a line. Label the resulting diagram with the values of S, .95 in our example, usually referred to as the threshold similarity. At this level all other items are unlinked and therefore not shown.

Search the similarity table for the next highest value, and draw the second linkage diagram. Do this by taking an exact copy of the first, alter the threshold value, S, to .93, and add the new link(s) to the diagram. Several situations may arise.

1. If the new link is between two items formerly unlinked, as in fig. 8.12a, place two labeled points on the diagram in arbitrary positions and

	Object Numbers						
	7	6	5	4	3	2	1
1	.82	.77	.78	.75	.90	.95	1.0
2	.80	.76	.79	.74	.93	1.0	
Object 3	.80	.83	.79	.80	1.0		
Numbers 4	.90	.92	.95	1.0			
5	.92	.85	1.0				
6	.85	1.0					
7	1.0						

Table of Similarity Values, S_{IJ}

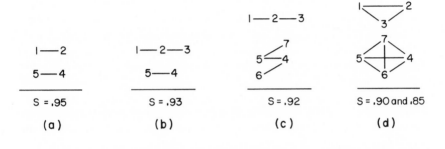

(a) S = .95 (b) S = .93 (c) S = .92 (d) S = .90 and .85

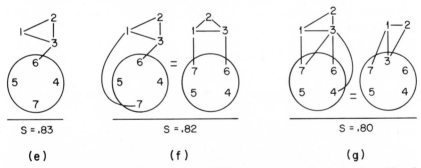

(e) S = .83 (f) S = .82 (g) S = .80

Figure 8.12 Steps in drawing a series of linkage groups on the basis of a table of similarity values.

join them by a line. The effect of the diagram is best if the user avoids cross-ing lines where possible. The new link is the start of a second cluster.

2. If the new link is between an item already linked and an item not linked, as in fig. 8.12b, merely add one new labeled point and join it by a line to the point already linked, thus increasing the size of the cluster.

In some cases more than one item will join (fig. 8.12c). As the process is repeated, two further cases arise at later levels.

3. Both items with the next highest similarity value are already members of different clusters. In this case join the two labeled points (already on the previous diagram) by a line, so joining two clusters together (fig. 8.12e).

4. Both items with the next highest similarity are already members of the same cluster, as in fig. 8.12d. In this case, again join the two labeled points by a line. This is called an **internal link:** it makes no difference to cluster membership but aids the diagram by contributing visual representation of the internal connectedness of the cluster.

In practice there is no need to draw separate diagrams for every level at which links occur (fig. 8.12d represents S values of .90 and .85). The question of when to stop making additions to one diagram and start making additions to a copy is simply one of which intermediate stages the user wishes to preserve. Clearly, if the process were carried to completion on a single diagram, all items would eventually be interlinked in every study.

Although in the instructions given above the user was encouraged to keep links approximately equal in length and to avoid intersections, as more points and links are added this becomes progressively more difficult and finally impossible. For homogeneous clusters, crossing internal links are unavoidable. As the diagram becomes more complex, the user will realize that he is producing a compressed, flat diagram from what is conceptually a multidimensional structure. Sometimes reversing, inverting, or twisting parts of the linkage diagram will simplify it, even cut link lengths or reduce intersections (fig. 8.12e shows an inversion that reduces a long link between 3 and 6).

For large and homogeneous clusters even the most skillfully untangled linkage diagram will become unmanageably complicated so that some simplification is needed. One such simplification is the **circle-criterion.** Any set of items within which each is linked to three or more others is substituted for by a circle. Fig. 8.13 gives the rules for drawing circles. Labeled points representing the items are placed round the circumference and only links with items outside are now recorded. As the clustering proceeds—that is, as the threshold similarity is reduced—more and more pairs of items reach the threshold and are linked. Consequently the number of items in the circles increases and eventually the circles coalesce.

The resulting diagrams are frequently a mixture of circles and linked points. The final effect is visually meaningful and carries much more information than the dendrogram. All of the information on a dendrogram is depicted in a linkage diagram, but details of the clustering process are added (see fig. 8.14). For instance at a given threshold similarity, linkage diagrams show which items are linked to others and which belong to homogeneous parts of clusters represented by circles. Chains, intermediate chaining between homogeneous clusters, chains of outliers, and solitary outliers are all

A. To start a circle: each object included must be linked to 3 or more other objects. This implies there are 4 or more objects or one of these patterns:

B. To join a circle: object must form 3 or more links with members, as object A, and then B above.

C. To fuse two circles: each object in the newly formed circle must have links with at least 3 other members, as in both cases above.

Figure 8.13 Criteria for drawing circles to indicate closely linked objects.

clearly recognizable. Whereas each cluster is represented as a stem on a dendrogram, on a linkage diagram such a cluster has its internal structure displayed. It may be shown to contain one or more homogeneous parts linked by crosslinks or intermediates, possibly surrounded by outliers or outlying chains. At a node in the dendrogram, two clusters are shown as fusing together. The linkage diagram gives further information. It shows which pair of items, one from each cluster, has caused them to fuse together. This in turn shows whether it is the homogeneous cores of the two clusters which are similar or whether a chain of intermediates has caused them to fuse. In fig. 8.14 the linkage diagrams show three quite different ways in which the clusters α, β and γ might have produced an identical dendrogram.

Until recently the principal disadvantage of linkage diagrams has been the need to draw them by hand. For large numbers of items this is laborious and leads to errors. However, experience with output from Rogers' GRAPH program suggests that users quickly become skilled at drawing,

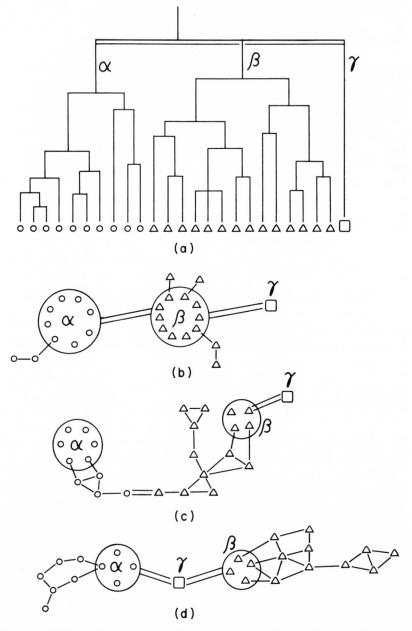

Figure 8.14 Three different linkage diagrams (b, c, and d) showing three of the possible fusions leading to clusters α, β, and γ, any of which could be depicted by the nodes in the dendrogram (a).

untangling, and crosschecking for human errors. Using a copier to reproduce the linkage diagram at the previous level as the basis of the next diagram eliminates one source of human error.

The problem for computer implementation is the arbitrary spacing and need for tidying and untangling. Some progress has been made recently in interactive untangling of linkage diagrams. If the network is displayed on the video screen of an interactive terminal, the user can give instructions for repositioning and untangling. Although the decisions are made by the user, the program can at least check that no errors, such as mislaid links or items, occur in the altered network. Examples of linkage diagrams applied to several cases of actual taxonomic work are given for *Manihot* in Rogers and Fleming (1973), for *Crotalaria* in Bisby (1973), for the genera of Chrysobalanaceae in Prance, Rogers, and White (1969), for *Salmo* in Legendre, Schreck, and Behnke (1972), for Genisteae in Bisby (1981) and Harris and Bisby (1980), for *Vicia* specimens in Birch, Tithecott, and Bisby (1985).

Minimum Spanning Trees

If items studied are depicted in a geometric model, the minimum spanning tree is the sequence of links between items that couples all the items together in such a way that the total length is minimized. It is a tree in the terminology of graph theory, in that it consists of lines joining points with forks, loose ends, and no closed loops (Prim 1957, Ross 1969). Notice that it uses precisely the same information as does the single-link clustering method (see fig. 8.15). It can be drawn either with links of equal length or with each link drawn of length equal to the distance between the points in the geometric model. Two useful aspects of a pattern are shown well by a minimum spanning tree. It shows for each item (1) which is its closest neighbor, and (2) where there are gaps between clusters, which is the closest item of those outside the cluster. The convention is to draw junctions in the tree as right-angles, but clearly any angle can be used; some other angle must be used where five or more links join. Examples based on serological data can be seen in Cristofolini and Chiapella (1977). The usual practice is for the computer program to print the minimum spanning tree sequence of items and the similarities between them on the line printer from which the line diagram can easily be drawn by hand.

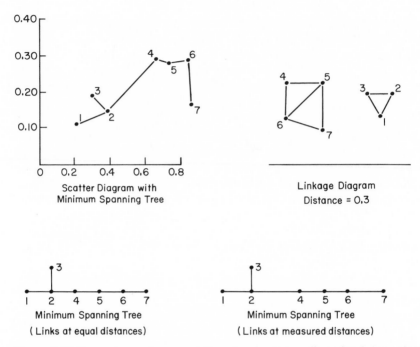

Figure 8.15 Comparisons of representations for simple two-dimensional data using a scatter diagram, minimum spanning trees drawn in two styles, and a linkage diagram.

Combined Tactics

Some of the most successful pattern display diagrams combine information from two different pattern analyses. Some examples are:

Scatter Diagram with Minimum Spanning Tree

This, probably the most widely used combination diagram, was introduced by Gower and Ross (1969). Points are drawn on a 2-D scatter diagram in the normal way, but these are linked by lines in the sequence of the minimum spanning tree. The outcome is particularly useful with scatter diagrams of PCA or PCO results, where it overcomes the difficulty of knowing whether points close in the scatter diagram are close in all dimensions (Ross 1969). The method is most effective with relatively small numbers of items, as in

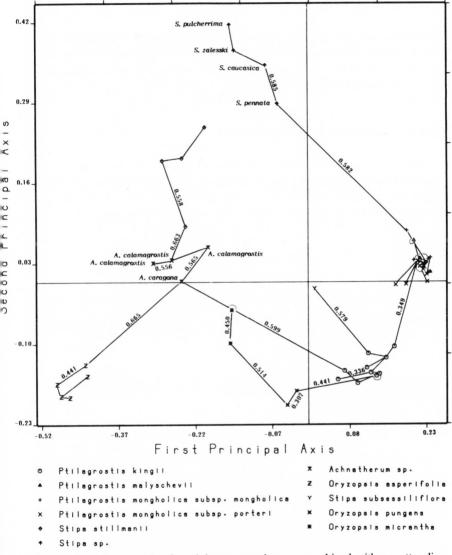

Second Principal Axis

First Principal Axis

○ Ptilagrostis kingii
▲ Ptilagrostis malyschevii
+ Ptilagrostis mongholica subsp. mongholica
✕ Ptilagrostis mongholica subsp. porteri
◆ Stipa stillmanii
✦ Stipa sp.

✗ Achnatherum sp.
Z Oryzopsis asperifolia
Y Stipa subsessiliflora
✕ Oryzopsis pungens
✱ Oryzopsis micrantha

Figure 8.16 An example of a minimum spanning tree combined with a scatter diagram to illustrate relationships among a small number of grasses. *(From Barkworth 1983. Fig. 10).*

Barkworth's (1983) study of *Ptilagrostis* grasses in fig. 8.16. As more are added dense regions become clustered and parts of the information are obscured. It can also be used on three-dimensional plots for PCA as shown by Schnell (1970) in his work on gulls.

Dendrogram and Order on Principal Axis

In dendrograms the ordering of stems which fuse at one node of the hierarchy is normally arbitrary. Further information can be incorporated by making this order follow some computed order, such as the order of the items on the principal axis of a PCA or PCO.

Scatter Diagram and Cluster Contours

A dendrogram gives no spatial information about clusters even though it may be based on an analysis of a geometric model. Spatial information can sometimes be effectively achieved by delimiting clusters by contour-like lines on a 2-D scatter diagram. Moss (1967) gives some good examples.

Suggested Reading

Sneath, P.H.A. and R.R. Sokal 1973. *Numerical Taxonomy*. San Francisco: W.H. Freeman.

Chapter 9

Identification

We have seen in chapter 2 how identification is an important process for those who wish to use the biological information system. Merely determining the correct species name for an organism is of little value, but in most practical contexts the name is needed to connect the specimens of interest with the appropriate information in the system. Consequently, providing a means of identification, an **identification aid,** is part of a taxonomist's job; and the most common type of aid, identification keys, are indeed provided in most monographs, Faunas and Floras.

Anyone who has tried to identify many organisms will know that the process is frequently troublesome. The principal difficulty is of course that many taxa are genuinely difficult to distinguish, but other problems are important too: the person making an identification is usually unfamiliar with descriptive features of the taxon in question; the organism may reveal different diagnostic features at different stages of its life history; and identification aids are of variable quality. A common complaint is that identification aids produced by taxonomists are pitched for use by other taxonomists, both in the use of technical terms and in the assumption that a preserved specimen is under examination. As a result secondary publications by skilled field biologists have sometimes proved more accessible to nontaxonomists.

Identification is the act of deciding to which part of a given classification a specimen belongs. A person who examines, say, a specimen tree and locates its position within the family Leguminosae is said to have identified it to the family level. Most commonly the identification will be continued to the species level—perhaps, for example, discovering that the tree is a cultivated Redbud or Judas Tree, *Cercis canadense* in North America or *Cercis siliquastrum* in Europe.

Identification and classification are often confused. These two words are used in biological taxonomy to mean quite different processes.

The process of classification leads to an arrangement, system, or grouping of organisms also known as a classification. Logically this must precede both the nomenclatural process (giving names to taxa in the classification) and identification (deciding where a specimen belongs in the classification and finding the name of the taxon to which it belongs). Unfortunately these words 'identification' and 'classification' differ from their usage in other disciplines. What taxonomists call identification is called diagnosis in medicine, classification or discrimination in statistics, and naming in common parlance. We should also note that someone who thinks the sole purpose of taxonomy is to identify specimens in fact neglects the caveat that a classification and nomenclature are needed before this is possible.

Generally, there are just two methods of identifying organisms: by matching and by elimination. Matching depends on the availability of a comprehensive set of specimens or illustrations and keen observation. Elimination depends on knowing what are the possibilities and how to eliminate them on the basis of identificatory characteristics. The success of the matching process, whether used by a nonbiologist with an illustrated field guide or by a professional taxonomist searching specimens in the British Museum, comes largely from the relative ease with which humans can spot or compare a whole assemblage of visual characteristics. In comparison the use of a written or coded description is often clumsy. The attraction of elimination methods lies in the precise objectivity possible once the technicalities are mastered and once the information needed is available.

The last decade has seen tremendous advances in two areas: in the use of computers in preparing identification aids and in the quality of identification aids printed in field guides for use by nontaxonomists. We mention briefly some of the advances in field guides below. The use of computers is the subject of the remainder of this chapter: in it we distinguish the computer preparation of hand held identification aids from the use of computers themselves to perform the identification. An excellent survey of identification techniques and modern developments is given by Pankhurst (1978).

Advances in Field Guides

The principal development here has been simply the enormous improvement in the quality of printed illustrations. Color photographs have played their part (see Polunin and Smythies 1973), but the major advance is with color paintings. Books like Keble-Martin's *The Concise British Flora in Colour* (1965), Schauer's *A Field Guide to the Wild Flowers of Britain and Europe* (1982), Chinnery's *A Field Guide to the Insects of Britain and North Europe*

(1973), Higgins and Riley's *A Field Guide to the Butterflies of Britain and Europe* (1970), Covell's *A Field Guide to the Moths of Eastern North America* (1984), and Burt's *A Field Guide to the Mammals of America North of Mexico,* 3rd Edition (1976) combine high-quality illustrations of very large numbers of organisms with authoritative classification and nomenclature. Most notable among these illustrated field guides are the series produced by Peterson in the USA and by Collins in Europe.

A second publishing advance is in the series of identification handbooks produced by organizations such as the Botanical Society of the British Isles, The Linnean Society, and the Royal Entomological Society. Despite their conventional monographic layout—keys, descriptions, line drawings—they have been carefully produced for use by nontaxonomists. The terms are explained or illustrated, the keys are tried and simplified where possible, and the illustrations are very carefully selected and executed. Lousley and Kent's *Docks and Knotweeds of the British Isles* (1981) is an example containing black and white drawings of quite exceptional clarity and beauty.

Computer-Produced Identification Aids

Computer-produced identification aids use the elimination method of identification. This method assumes that the specimen belongs to one of a given list of taxa, that the P taxa have names and that all or part of a list of n identification characters have been observed and recorded for each of the taxa. Provided that there is sufficient discriminating data in the $n \times P$ data table, it is possible, by examining the identificatory characters in turn, to eliminate progressively more of the P taxa until P-1 have been eliminated and the remaining taxon is the correct identification. Identification keys, also known as **diagnostic, dichotomous,** or **single entry keys,** contain a single route by which progressive elimination leads to the identification. First one answers the first pair of leads, then the second in the appropriate branch, then the third, and so on as illustrated in chapter 2. **Polyclaves** or **multiple entry keys,** by contrast, contain a variety of routes and the user chooses at each stage the next identificatory character to be examined. Examples are given below.

Keys

Morse (1968, 1971, 1974), Pankhurst (1970, 1971), Hall (1970, 1973, 1975), Payne (1975) and Dallwitz (1974) have produced computer programs which construct keys from a comparative data table. There is a basic similarity in the procedures used by all of them: a recursive procedure divides the items into progressively smaller subsets. Each division is into two or more subsets possessing different character states for a chosen best character. These two or more character states form the contrasting leads of a key. The process is repeated until all items are keyed out or, where there is insufficient data for complete resolution, until all items have been discriminated as far as the data allow.

The procedures choose the best available character for each division, employing a measure of usefulness which incorporates information on:

1. Character weighting by ease-of-use or cost-of-use values;
2. Character weighting by the extent to which the character will produce a balanced division;
3. Taxon weighting by frequency of occurrence.

Ease-of-use or cost-of-use values are supplied by the user. The values are incorporated into the measure of usefulness so that they bias the choice of character toward those which are considered easiest or cheapest to use. Payne (1975) incorporates too the possibility that later parts of the key may utilize different character states or value ranges of the same characters used at earlier stages, thus saving on cost-of-use, since the character must already have been observed.

Balanced divisions are preferred in choosing between characters, all else being equal with regard to character and taxon weights. The result is a balanced key in which all branches contain similar numbers of leads. In this case nearly the same number of leads will be answered for all taxa to which specimens might be identified. It does not of course change the total number of leads. By contrast an unbalanced key allows some taxa to be identified by answering only a few leads while others require many leads. If the probability of error in answering any lead is fixed, then the chances of making a correct identification decline markedly when one answers many leads—a real disadvantage in addition to the extra labor involved (see fig. 9.1).

Taxon weighting is used when it is known that some taxa will be encountered more frequently than others by users of the key. It provides a bias toward characters which will enable the taxa in question to be keyed out near the beginning of the key. This inevitably makes the key unbalanced, and it raises the number of leads needed to identify a taxon of more rare occurrence; but it does reduce the labor and increases the accuracy of identification of common taxa.

Other features which can be varied in one or other of the programs are:

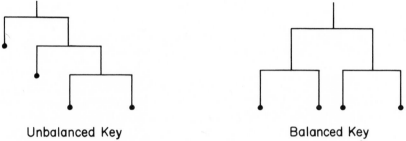

Unbalanced Key   Balanced Key

Figure 9.1 Diagrams of unbalanced and balanced keys.

1. Style of print;
2. Use of accessory characters;
3. Treatment of unknown or variable values of a character for a taxon;
4. Treatment of taxa not discriminated by the data supplied;
5. Treatment of multistate and continuous characters.

Pankhurst (1971) provides the two conventional styles for print-ing keys, the indented key and the bracketed key, as illustrated in chapter 2. Payne, Walton, and Barnett (1974) have experimented with other styles, particularly for binary characters where the character states have repetitive descriptions such as positive/negative. They provide a condensed format, which is illustrated in fig. 9.2.

Accessory characters at any one decision point are additional characters that, if used, would give the same (or nearly the same) division as the character chosen. In a practical sense they are particularly useful if they happen to have conditions of use which complement the character that was selected. For instance, where a flower character is selected in a botani-cal key, accessory vegetative characters can be very helpful for use with specimens lacking flowers.

The treatment of variable values provides one of the most subtle differences between the programs. Variable values occur where two or more character states are recorded in the same taxon. Clearly, if the program can devise a key which can avoid the use of variable values, then this should be done. Payne (1975) calls this stage 1 selection. Stage 2 uses variable values and, as a consequence, the taxa bearing these values key out in two different places in the key. Stage 3 selection involves use of prior probabilities re-corded within the variable taxa, so giving leads in which the identification is "probably A but possibly B."

Unknown or missing values in the original data table are a nui-sance. If at some stage a character with a missing value is used, the taxa for which it is missing have to be entered in each of the subordinate branches, thus lengthening the key unnecessarily.

All of the programs can make partial keys where the data pro-

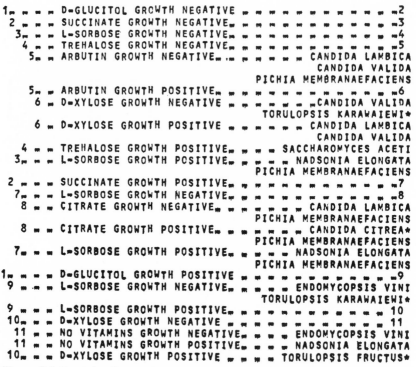

Figure 9.2 Example of a condensed format for a key using binary characters for bacteria. *(From Payne, Walton and Barnett 1974. Fig. 1 d).*

vided do not discriminate between all of the taxa. For computation purposes groups of such taxa can be treated as a single taxon with two or more names. A more difficult context is where the user has chosen between a partial but reliable key and a complete but probabilistic or even unreliable key which uses variable values. In practice, this is often resolved by introducing more data during a repeat run of the program.

The treatment of multistate and continuous characters varies between the programs. Things to look for are whether continuous characters are subdivided by arbitrary subdivisions or by natural gaps detected by either the user or the program. Some programs allow different splits of the same character to be used at different stages in the key. Payne's program allows one to choose deliberately to do this, thus getting extra value from a single set of observations.

The principal justification for using the computer to produce keys is that it saves labor, and that the keys are relatively error-free. Whether the keys produced are published directly, as in Webster (1969, 1970) and Barnett and Pankhurst (1974), or are used as experimental suggestions for a final edited version, as in Watson and Milne (1972), the computer has taken a large share of the work. However, this may produce a real saving in time only if the data table is relatively easily obtained. Thus, where data have already been tabulated for a monographic study it may just be a matter of selecting and editing a subset of clearcut, easily observed characters from a larger set already observed.

If the operation is started from scratch, then forming a complete data table can represent an unreasonable burden. The conventional way of making a key need not involve the observation of all characters for all taxa. In theory only one character (that used in the first decision) need be observed for all taxa, but in practice the search for such a character involves other observations too. It is true that many of the keymaking programs will also make a key from a table with much missing data, but the true value of the programs in selecting the best structure is realized only if there are sufficient data to offer a choice of structure.

The time needed to produce a complete data table may be particularly well spent if a number of slightly different keys are needed for approximately the same taxa. If editing facilities allow deletion of data rows or columns, or masking facilities allow them to be masked out, some of the taxa (columns in the data table) or some of the characters (rows in the data table) can be removed before the program is run. So, for instance in a worldwide monographic study of a plant group (such as the Vicieae Database Project, Adey et al. 1984) deleting or masking taxa will be used to provide special keys for the plants found in different geographical areas, and deleting or masking characters to provide special keys for identification of seeds, fruits, or vegetative fragments.

Polyclaves

Polyclaves, or multiple-entry keys, work on the same principle as dichotomous keys i.e., by successively specifying which state of a key character is present for the unidentified organism and eliminating other taxa until just one is left. The difference is that the user chooses which key character to use first, and at each subsequent step; herein lies the advantage of the polyclave. How often has the user of a key been frustrated by being unable to make the observation needed to answer the first or early leads in a dichotomous key? We have fruits and vegetative material available, but the important early leads in the dichotomous key involve characters of the flower! Try for instance identifying one of the *Quercus* (oak) species from an autumnal sample of leaves, twigs, and acorns. Undoubtedly the specimen should be identifiable to the species level, starting with no knowledge that it is even an oak, but doing this is almost impossible using the keys in the *New Britton and Brown Illustrated Flora* (Gleason 1968) or the *Flora of the British Isles* (Clapham, Tutin, and Warburg 1962) because the key to families depends on characters of the flower. By contrast, a polyclave would allow the user to choose which character to answer first.

A difficulty with polyclaves lies in their physical arrangement. They are, in a sense, multidimensional and hence hard to arrange on the printed page. The most common form, until recently, has been a pack of edge-punched cards. Fig. 9.3 is an example, from a microscope key to hardwoods prepared for the Forest Products Research Laboratory.

Each of the numbered holes around the edge is used as a binary character with the name of the character printed adjacent on the card. A ticket punch is used to clip out the edge of some of the holes. If the hole is clipped, plus is signified for that character; if it is not, then minus is signified. (Or vice versa!) To make an identification the user decides on which character to observe first, and then pokes a knitting needle through the appropriate hole in the complete pack of cards. Then the user lifts the needle and shakes the pack. All cards corresponding to species with unpunched holes for the character stay on the needle, and all with punched holes fall off. If the punched holes signify plus and the character in question is present on the specimen, the user takes the subset of cards left on the needle, and repeats the whole process for another character. Each card corresponds to one species, and when the elimination process has gone down to one card, the specimen is identified by the name of the species on that card. The problems are mechanical: how to avoid crumpling or spoiling the several hundred cards in use; how to be sure that all fall out of the pack when they are not on the needle; how to store the pack; and how to reproduce multiple copies of the cards properly printed and punched.

A more recent version of the polyclave uses 80-column punched cards of the kind once much used to enter data and instructions into computers. Each card represents one character state and has holes punched at numbered sites within the card rather than on the edges. The user selects cards appropriate for the character states he observes on the specimen. As a pack of selected cards builds up it is held up to the light to see how many holes are punched right through the pack. The identification is made when there is just one hole with the light showing through. The number of this hole tells the user to which taxon on the accompanying list of taxa the specimen belongs. Because of their unclipped edges and the elimination of the needle selection cards of this type are less prone to damage in use. However, punching machines for printing and punching holes on such cards— once widely available—are now disappearing.

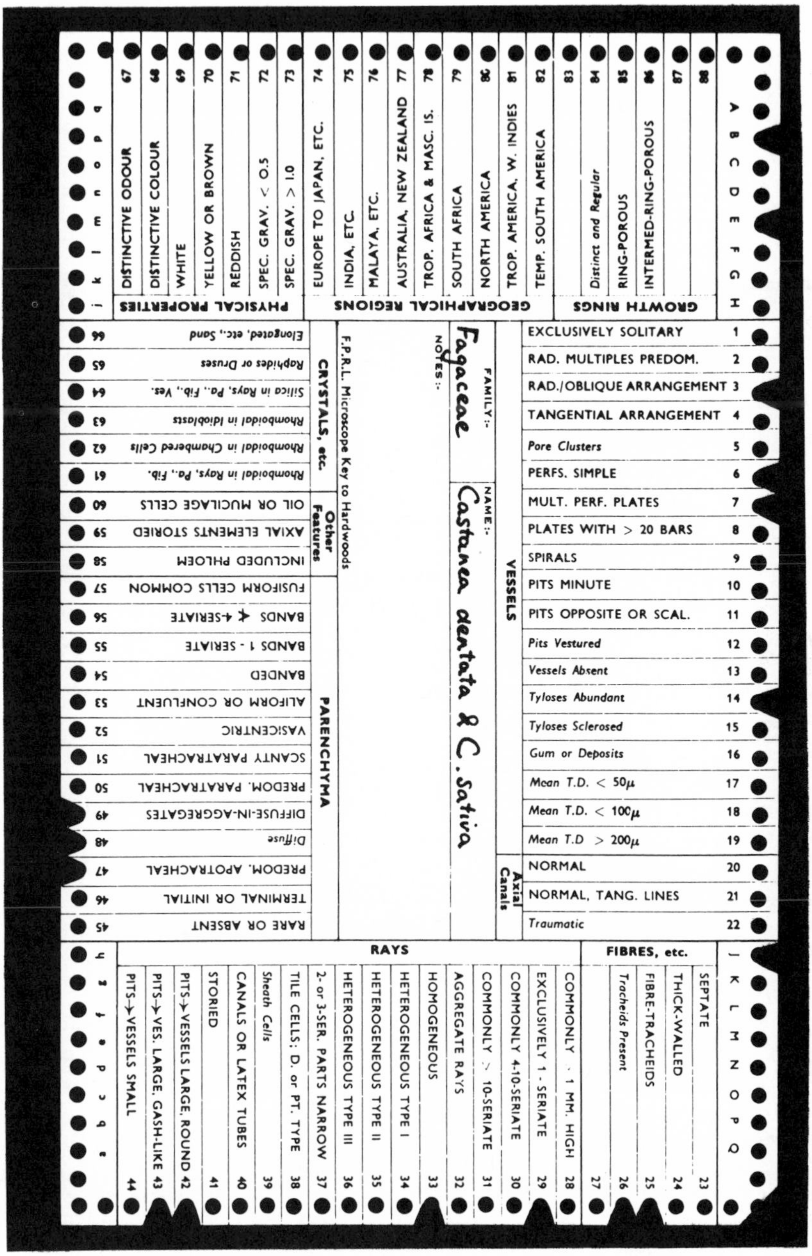

Figure 9.3 Example of an edge-punched card key. *(From A Microscope Key to Hardwoods (Anonymous). Published by HMSO, London. Used by permission.)*

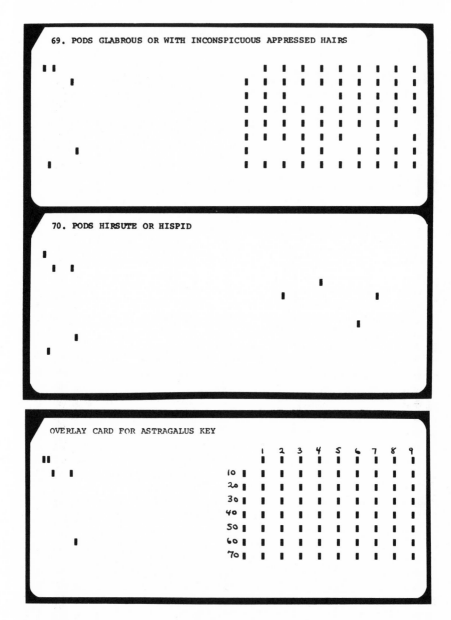

Figure 9.4 A punched card key for the genus *Astragalus.* (above) When these 2 character cards are superimposed only 2 holes remain uncovered. Use of the overlay indicates that these are #32 and # 37, which correspond to *A. gilviflorus* and *A. hyalinus* on the species list. (next page) An example of the species cards in which punched holes correspond to presence of numbered characters (attributes) on an accompanying list. *(From Weber 1975. Cards kindly provided by W.A. Weber).*

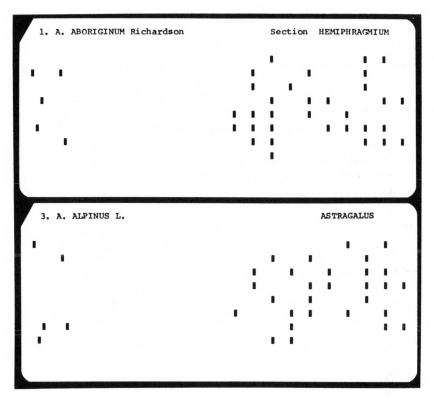

It is of course possible to design a polyclave by hand and just have the product put onto cards as in the "Punched card key to the Dicot families of South India" (Saldanha and Rao 1975) and the "Polyclave Key to the Species of *Astragalus* in Colorado" (Weber 1975), which is illustrated in fig. 9.4.

Pankhurst and Aitchison (1975), Morse (1974), and Johnston (1980) have produced computer programs which, when a multivariate data table is supplied, perform much of the logical organization involved in making the polyclave and also automatically produce the cards at the end.

The Johnston (1980) program illustrates the similarity between using a polyclave and searching a database. In addition to the character cards the program can produce a set of cards where each card bears the name of a taxon and numbered holes are punched to correspond to characters like those shown in fig. 9.4(b) of the *Astragalus* key. Thus, by looking at the card for a species one can read off the characters that are found in that species. If the two packs of cards—the polyclave with a card per character state and the data set with a card per species—are used at the same time, the user can both identify a specimen using the polyclave and verify that this identification is correct by checking from the data set cards the other characters that it should possess. The relationship between a polyclave and a database should now be clear. These two packs of cards contain the same data. In the one case the data are printed on cards by rows—the character cards of the polyclave—and in the other by columns as the species cards of the data set.

An interactive database such as those described in chapter 11 can be used in the same way as a polyclave key to eliminate all but one possibility by successive stages. The result can then be checked by reading off characters or descriptions contained in the database.

Practical inconvenience is still the principal reason why most key identification is with conventional single-entry keys, not polyclaves. An innovation that has brought a kind of polyclave into the mass-production market with field guides is well developed in *The Shrub Identification Book* (Symonds, 1963). In this each species is illustrated: its flowers in the flower section, its fruits in another section, and so on. The user matches the specimen with the illustrations, but repeats this for as many organs as possible in different parts of the book. On each occasion a note or list of the possibilities is made. These notes or short-lists are placed side-by-side and in most cases only one species is common to all of the lists. This is the correct identification. These tentative identifications can be checked by looking at the final section of the book in which the illustrations are repeated, this time with all organs of the same species grouped together.

Identification by Computer

There are a number of computer programs that can be used to identify specimens. The user supplies observed data for the specimen and the program identifies the specimen's position in a classification, which has already been supplied. The programs use elimination techniques, matching techniques, or both in combined tactics.

Elimination: Keys and Polyclaves

Some of the earliest computer-identification means were effectively computer-operated single-entry keys (Boughey, Bridges, and Ikeda 1968). The principle was just the same as for a key: users choose which of the leads presented is correct and this takes them on to the next appropriate set of leads. Now the set of leads is printed by the computer (usually at a terminal), the user indicates which lead is correct either by typing a response or by pointing with a light pen or cursor, then the computer prints the next set of leads. As Morse (1975) points out, such a program normally has no advantage over a printed key, unless it can be used to edit or adapt the key (as suggested in Sokal and Sneath 1966, and Van Dam and Rice 1971) or it can use computer graphics to illustrate the leads. An example of using graphics is the fungal key designed by Kendrick (1972). This presents the user with a pair of illustrations between which to choose. The leads are ordered from gross characters to small-scale characters, and as the procedure progresses, the illustrations build up successively more and more detailed magnified diagrams of parts of the organism. Pankhurst (personal communication) has now taken up Morse's other suggestion and produced an interactive program which allows the user to edit a single-entry key.

In contrast, most recent computer identification programs are at least approximately based on a polyclave, but with other features added. While a computer terminal is not as cheap or portable as a deck of cards, in other ways the computer implementation can be ideal. A variety of programs, of which the best known are those by Morse (1974) and Pankhurst and Aitchison (1975), involve sequential elimination and include an extensive interactive dialogue between the user and the computer. At each cycle the user chooses another character to observe, the computer supplies the leads or states of that character, and the user responds by choosing the one that fits the specimen. A sequence of these cycles is terminated when only one taxon is left uneliminated and the identification is made. Some programs allow the sequence to be interrupted for the user to find out how many possible taxa remain uneliminated, and some provide suggestions as to which characters or leads, if answered next, will discriminate between those that remain. The example given below (fig. 9.5) is from a student's very simple interactive polyclave program at Southampton University. A more sophisticated example is given in Pankhurst (1978 pp. 70, 71).

Matching Methods

There are some examples of computer identification programs which use the matching process, albeit matching of coded descriptions of the taxa. One

```
1   EMERGENCE OF FIRST LEAVES AS A WHORL.                                    26.GALIUM APARINE
    EMERGENCE OF FIRST LEAVES UNEQUAL.                                                        2
    EMERGENCE OF FIRST LEAVES OPPOSITE TO EACH OTHER.                                        25

2   FIRST PART EMERGING FROM GROUND IS LEAVES SIMILAR TO FOLLOWING LEAVES.     65.VICIA HIRSUTA
    FIRST PART EMERGING FROM GROUND IS CYLINDRICAL GRASS COLEOPTILE; NOT LEAF LIKE.           3
    FIRST PART EMERGING FROM GROUND IS A PAIR OF COTYLEDONS, UNLIKE FOLLOWING LEAVES.         7

3   HAIRS ON FIRST LEAF SPARSE.                                                               4
    HAIRS ON FIRST LEAF ABSENT.                                                               5

4   LENGTH OF LIGULE UNDER 1MM LONG.                                        1.AGROPYRON REPENS
    LENGTH OF LIGULE OVER 1MM LONG.                                             9.AVENA FATUA
                                                                        10.AVENA LUCOVICIANA
```

Figure 9.5 Example of the printout from a computer polyclave identification program.

uses identificatory characters that are mostly invariant within the taxa. The other allows the use of characters that vary within as well as between the *P* taxa. In the latter the data are thought of as representing a multidimensional space containing regions that are within the known boundaries of one of the *P* taxa and others that are outside them.

In both techniques all *n* characters are recorded for the specimen to be identified. In the first technique the description is compared with the *P* taxa descriptions. Either the program detects that it matches one exactly, a confident identification, or the program calculates with which of the *P* taxa it matches in all but one or two characters or a specified allowance of mismatching. The second technique (Gyllenberg 1965; Niemala and Gyllenberg 1975) gives a confident identification if the specimen falls within a taxon boundary. If the specimen falls outside the boundary, it provides a measure of the shortest—and sometimes second, third, and fourth shortest—distances from the specimen to a boundary (fig. 9.6).

We have now implied that possibly a specimen might *not* exactly fit any of the known descriptions of taxa. This can happen, surprisingly frequently, if the specimen does not belong to the taxa for which the iden-

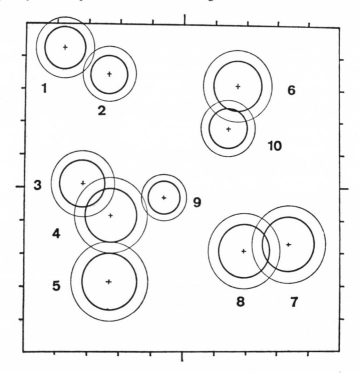

Figure 9.6 Diagram of taxon boundary relationships in an identification by a matching method. *(From Niemala and Gyllenberg 1975, Fig. 10 p. 145).*

tification aid is intended. An example might be an Englishman identifying *Cercis siliquastrum* in *The Flora of the British Isles*. It is not included because it is an alien plant, but the innocent person who looks it up cannot tell this just by looking at the plant.

Another example is the deliberate use of taxonomic works intended for one well-known area of the world by botanists working in a poorly known area of similar flora. Northern Morocco is a case in point: Moroccan taxonomic coverage is poor and the best work for Northern Morocco has been *Flora Europea*. However, there are plants in Morocco which are not in that book.

A different problem is the occurrence of unexpected, extreme, intermediate, or hybrid forms in identifications at or below the species level in very variable groups. The apomictic species of *Rubus* or the complicated local races of *Silene* species are examples where identification, particularly of single specimens, are probabilistic not deterministic; the variability and complexity of the pattern make identification approximate rather than exact. In these cases it is more useful to know to which taxon it most nearly fits than to discover simply that it is an extreme form which cannot be identified for certain!

An even more difficult problem is found in the low-level taxonomic pattern of some groups. This is when the range of variation in similar taxa actually overlap, even in the most discriminatory characters. Statistical methods for dealing with this problem, that of discrimination rather than of conclusive identification, are well-known but will not be dealt with here. Discriminant Function and Canonical Analysis, which have been discussed in chapters 5 and 7, are two tools for such discrimination.

Conclusions

There is now an excellent range of programs available for making and editing keys, making polyclaves, and for performing computer identifications. A naïve observer might be puzzled then as to why these techniques are in fact getting rather little use at present. We suggest a number of reasons.

No amount of sophisticated optimization of keys or computer routines will make for quick, easy identification if the identificatory characters are not themselves clearcut characters, both in the sense that they have clearly distinguishable states which can be observed by the novice, and in the sense that these characters are consistently different for the taxa in question. To a large extent, meeting these criteria requires skilled, knowledgeable observation of the plants by a human and is unaffected by the subsequent computer work. A further problem here is that characters may be clearcut (in either sense) among some of the taxa but variable or indistinct in others. In a key produced by conventional methods this character might be used, but only in one particular branch of the key. The computer programs, however, require a character weighting factor to be applied over the whole set of taxa. We would expect techniques, such as that for key-editing, which allow a more detailed collaborative interaction between the taxonomist and computer, to help overcome these problems and eventually to lead to the regular use of computers in key-construction.

In addition, the advantages of using computer methods may be at their greatest in the production of identification aids for large groups. Yet it is here that constructing the necessary data table requires the largest amount of work. As yet, the direct communication of data sets between databases, taximetric analyses, and identification computing is in its infancy. However, once this becomes commonplace we may expect the use of identification computing to rise in connection with large floristic and monographic projects. In such cases a data table might be obtained ready-made and there might be many different kinds of identification aids needed—possibly for different subsets of the data table.

Finally, most identification aids are used very occasionally by a large variety of people. Yet one would expect large-scale resources to be invested in making identification more efficient only where a person or institution depends heavily on a particular identification program that will be used time and time again. Thus, we find particularly highly developed identification routines in bacterial identification laboratories such as those at the Public Health Laboratory in Maryland or the National Type Collection in London.

It is easy to make mistakes while using any identification aid. Consequently *all* identifications should be checked by reference to a description and *all* identification aids should be accompanied by descriptions for this purpose.

Suggested Readings

General

Pankhurst, R.J. 1978. *Biological Identification*. London: Arnold.
Pankhurst, R.J. ed. 1975. *Biological Identification with Computers*. London and New York: Academic Press.
Payne, R.W. and Preece, D.A. 1980. Identification keys and diagnostic tables—a review. *Journal of the Royal Statistical Society*, Series A, 143:253–282.

Chapter 10

Phylogeny and Cladistics

Acceptance of the Darwinian theory of evolution and knowledge of genetics in the early twentieth century led naturally to fundamental changes
in taxonomic concepts. The species was now a unit of evolution, with mutation, genetic variation, natural selection, and isolation the chief parameters
varying in time and space. The higher taxa of plants, animals, and micro-
organisms were now thought of as groups of species linked both within and
between the groups by a *phylogenetic history* of evolution and divergence
from common ancestors.

   Two important conceptual issues that arise from these changes
are first what is the reason for using a hierarchical structure in the classification? Secondly, should one arrange the classification to reflect phylogenetic
history of the organisms? Earlier in this century, taxonomists were uniformly evolutionist and optimistic on both issues—the hierarchical structure
was thought to be appropriate *because* all organisms arose from a single
ancestor, and taxa close on the phylogenetic tree could be brought together
so that the classification reflected the tree. Today these are issues for lively
debate but, we should warn, we stand strongly opposed to both of the above
views. [Readers may wish to read more widely, possibly from Cronquist
(1968), Sokal and Sneath (1963), Davis and Heywood (1963), Hennig (1966),
Cracraft and Eldridge (1979), Joysey and Friday (1982), Wiley (1981) before making up their minds.]

   The reasons for our position are the following: First, the hierarchical structure is suitable because of its utility to man as an information-
storage device. The variation patterns observed sometimes do and sometimes do not fit easily into a hierarchical description. Second, it is not at
present useful, or possible, to reflect the phylogenetic history in the classification, for two reasons: at present (1985) there is no certain method known
for deducing the phylogenetic history (it is unknowable), and even were the

history known, to include either its branching-pattern, that is, its **cladistic pattern,** or its time-scale into the classification would in many cases conflict with the classification's information property of storing the pattern of homogeneities and discontinuities found among the organisms today.

Both difficulties are nicely exemplified in the controversy of "the salmon, the lungfish, and the cow." One source of argument between Gardiner et al. (1979) and Halstead et al. (1979) is over the cladistic tree for the three animals. Gardiner holds that the character distribution shows the tetrapods sharing a more recent ancestry with the lungfish (see fig. 10.1a) while Halstead argues that the tetrapods arose from crossopterygian fishes such as the salmon, as illustrated in fig. 10.1b. A second debate centers on how to classify the three, even if one accepts that the tetrapods arose from

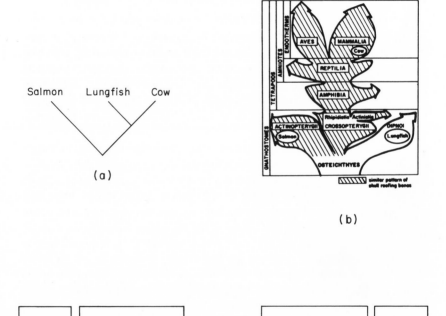

Figure 10.1 The Salmon, Lungfish and Cow controversy: a) and b) differing views on their descent; c) classification by supposed descent; d) classification by resemblance. *(b From Halstead et al. 1979, p. 176; a, c, d from Gardiner et al. 1979, p. 175.)*

the lungfish. Placing the lungfish and tetrapods in one class and the salmon in another (as in fig. 10.1c) would respect descent but violate the pattern of overall resemblances, while putting the cow in a class of its own (see fig. 10.1d) would respect the resemblances but violate the descent.

It is only realistic to recognize these limitations in our present ability to reconstruct phylogenies. With this in mind, the reader will understand why we have given considerable attention to phenetic pattern analysis methods in chapters 5 and 7, while in what follows we give only a brief outline of phylogenetic pattern analysis. It is also true that in general the 1960s saw a range of phenetic innovations by mathematicians, theoreticians, and programmers and that these took a decade for implementation, testing, and comparison by taxonomists. We have recently been through a similar period of exhilarating innovation in phylogenetic analysis, and there is still a constantly changing picture in developments and understanding.

We divide phylogenetic pattern analysis into four "schools"—those of Narrative Methods, Parsimony Methods, Cladism, and Compatibility Analysis. We comment on each, both as a candidate for adoption as a reliable method for deducing the phylogenetic tree and as a candidate for a classification method with information properties of use in the biological information service. Before discussing these schools we should make it clear how phylogenetic trees normally differ from phenetic classifications, and discuss possible "dual cases" where the two are identical.

Phenetics and Phylogeny

Phenetic patterns are patterns of overall resemblance and difference among organisms based on many heritable characteristics. Often there are discontinuities so that the pattern reveals groupings with differing ranges of variation within groups and varying degrees of difference between them. These patterns are usually observed among the organisms or fossil organisms at a given time. **Phylogenetic patterns** show how the phenetic pattern changes with time: they form a branching tree. In fig. 10.2 Sneath and Sokal (1973) show how, if we think of phenetic diversity as a two-dimensional pattern, phylogeny has a third dimension, time. A slice through such a three-dimensional tree gives a phenetic picture at a given time.

To enumerate the complete details of the phylogenetic pattern is impossible: we would need to know three components: (1) the branching sequence, known as the **cladistic tree** or **cladogram;** (2) the varying rates of evolution in **phyletic** (evolutionary) **lines;** (3) the actual age or timing of divergences. It is trying to deduce with confidence the first of these, the

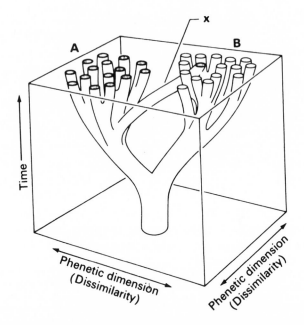

Figure 10.2 A diagram illustrating that phylogenetic pattern adds the time dimension to phenetic groupings based on overall similarity/dissimilarity. Phyletic line x has converged with other members of group B. *(From Sneath and Sokal 1973. Fig. 2–9, p. 57 of* Numerical Taxonomy—The Principles and Practice of Numerical Classification. *Copyright by W.H. Freeman and Co. All rights reserved.)*

cladogram, that has proved so fascinating and yet so elusive to recent taxonomists.

While the cladogram is of key importance in trying to reconstruct phylogenies, it is interesting to see how varied are the phylogenies that can be produced from the same cladogram by varying rates of change and timing. If the timing alone is changed, as in fig. 10.3, a phylogeny of, for example, essentially two lines may become one of four. If the rate of change and length of persistence is changed, the cladogram on the left of fig. 10.4 might suggest any of the very different trees on the right.

Fossil evidence is, perhaps surprisingly, of rather little use in determining cladograms or phylogenies. In the example of fig. 10.5, the fossils labeled A, B, C, with the relative ages shown, are of little help, as several phylogenies are possible.

Convergence causes the principal difficulty in deducing the cladogram. Convergence can be broken down into three effects: **parallel change** (the same change or mutation occurring at least twice, independently); **reversal** (a change being subsequently reversed); and **true conver-**

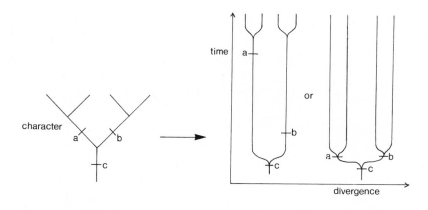

CLADOGRAM   PHYLOGENIES

Figure 10.3 Two very different phylogenies produced from a single cladogram by varying the timing.

gence or **mistaken homology** (resemblance occurring in nonhomologous characters). Once the last of these, true convergence, is detected it is no problem to avoid treating the characters involved as homologous, and by so doing remove any confusion that this may have been caused by the reconstruction problem. It is the first two, parallelism and reversals, known collectively as **homoplasy,** that cause the main problem. They give rise to the reticulate patterns of resemblance which are difficult to describe and classify in both phenetic and phylogenetic analyses.

Many biologists do not realize that cases where taxa have diverged at constant evolutionary rates appear to be very rare. However, in such rare cases the relationship between phenetic resemblance and the phylogenetic tree is simple. The phenetic distance will be a direct measure of evolutionary distance and time since divergence, as in fig. 10.6. Consequently the phenetic distance (and evolutionary distance) will have ultrametric (hierarchical) properties and we would expect a phylogenetic analysis and a hierarchical phenetic analysis to yield identical results: a clear hierarchy.

However in most real cases evolutionary rates are not uniform. In addition homoplasy (parallelism plus reversal) is present and so we must require any satisfactory phylogenetic or cladistic method to resolve them. In a sense we can use this ability to recognize and resolve homoplasies in the cladogram to distinguish cladistic methods from phenetic methods. Methods of analysis which cannot unravel the effects of homoplasy are likely to be basically phenetic.

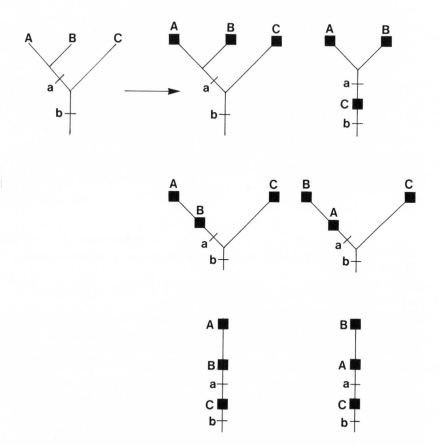

Figure 10.4 The cladogram on the left can correspond to any of the trees on the right. (A, B, and C are organisms, a and b are evolutionary steps such as the acquisition of a new character.)

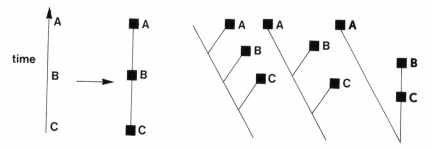

Figure 10.5 If A, B, and C represent fossils where C is known to be older than B and B older than A, then several trees are possible such as those on the right.

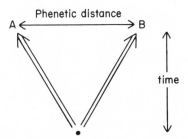

Figure 10.6 The relationship between time and phenetic distance if, and only if, the evolutionary rate is constant.

Narrative Methods

We include under narrative methods an ill-defined assortment of procedures notable perhaps, in retrospect, for the belief that the experienced taxonomist could sense the course that evolution had taken or the combinations of character states that represent milestones in its path without recourse to an explicit method of analysis.

Typically a world expert on a particular group would postulate a tree or balloon diagram, in which loosely defined phyletic lines are represented by lines or balloons as in figs. 10.7, 10.8, and 10.9. He or she would arrange major taxa in a simple directed tree structure, each with another present-day taxon as its antecedent and with general trends described within. The taxon whose line or balloon was at the base of the tree was either rooted in another taxon outside the study or was the only taxon in the study assumed to have arisen from extinct ancestral forms. The product was essentially an ordering of present-day major taxa into a simple, easy to remember scheme of the type referred to by Humphries and Richardson (1980) as a narrative.

The line of reasoning involved in sensing the narrative phylogeny was usually not made explicit, although there are exceptions such as Thorne (1976). This absence of explicit analysis, plus the fact that the author was more knowledgeable on comparative data on the group than anyone else, means that there was often no way in which the scheme could be tested or faulted. If a reasoning or defense was given, it sometimes took the form of following the progress of one or more ''important'' phylogenetic marker characters. These are chosen either because of their association with large functional or ecological differences between taxa, because of their strong association with the delimitation of the taxa so that they are also diagnostic characters, or because they show a sequence of changes with an intuitively

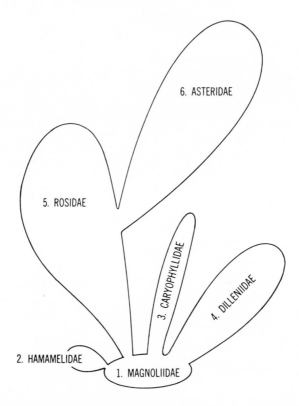

Figure 10.7 A balloon diagram illustrating relationships among Magnoliatae. *(From Cronquist 1968. Fig. 4.1, p. 129)*

appealing polarity of change, from simple to complex, from general to specific, and so on. The milestones are the appearance of new character-states or character-state combinations at the base of each balloon, and the balloon may be given an origin in another taxon either where some intermediary character-state or combination is found, or where the closest resemblance with the other taxon on many characters is observed.

The phylogenies produced by these traditional methods in fact only aim at a tree-like arrangement of the major taxa. The major taxa themselves have already been recognized on phenetic grounds—that is, on the basis of homogeneity within taxa or discontinuity between them based on overall resemblances. For instance, many of the families in Cronquist's scheme (such as the Ericaceae part of the Ericales in fig. 10.10) were recognized on the grounds of obvious overall resemblance long before the advent of Darwinism: they are phenetic groupings.

And further, even these treelike phylogenetic arrangements of

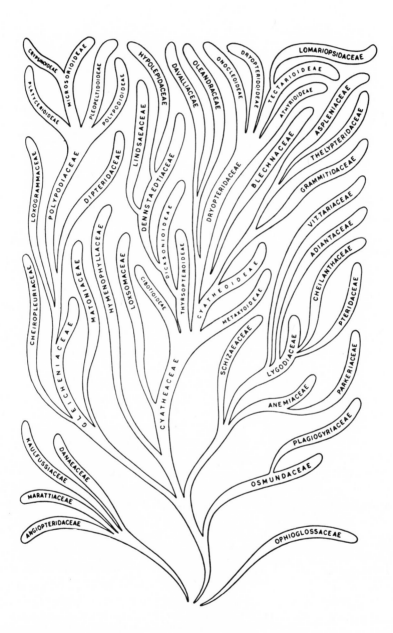

Figure 10.8 A diagram of narrative relationships in the ferns. *(From Nayar 1970. Fig. 1.)*

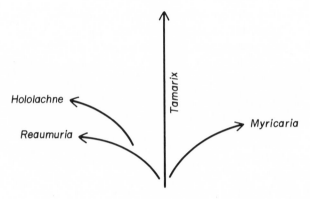

Figure 10.9 A line diagram showing relationships among genera of Tamaricaceae. *(From Baum 1978. Fig. 9, p. 17. Courtesy of Israel Academy of Sciences and Humanities).*

major taxa have been described as essentially phenetic (Humphries and Richardson 1980). This is because the tree arrangement has been constructed to some extent by linking together taxa according to their resemblance: that is, by phenetic methods and without our cladistic prerequisite, the ability to recognize and resolve homoplasies.

Another peculiarity, albeit one that is hard to overcome, is that the narrative methods assume that most of the major taxa arose from other present-day taxa. This is not so much because the authors believe the evolution actually happened very recently or that the evolution was from ancestors of these taxa identical to present forms, as because they have no explicit method for postulating the route by which later ancestors descended from postulated common ancestors. The problem of involving only present-day forms is also present in some of the explicit methods described below, but

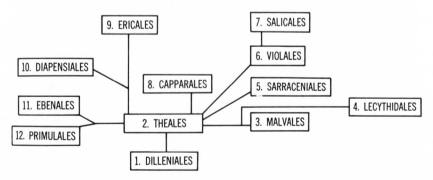

Figure 10.10 A narrative tree-like structure in which phenetic resemblances and phenetic groups are involved. (Groups are orders of Dilleniidae.) *(From Cronquist 1968. Fig. 4.6, p. 187).*

it is not an attractive feature. Some authors who adhere to essentially traditional methods have tried to remove this problem by depicting the scheme as the present-day cross-section of branches on a tree that diverges from common ancestors in the past, as in fig. 10.11 (Thorne 1976 and Dahlgren 1980). Similar diagrams however (such as the top surface in fig. 10.2) are used by Sneath and Sokal (1973) to explain that such a cross-section provides a purely phenetic picture.

So we must conclude that these traditional methods of narrative phylogenetic analysis are in practice unlikely to reveal the path actually taken by evolution. As methods of making classifications they seem to have worked quite well, and it can be argued that this is because of the phenetic element incorporated in the analysis.

We should not however dismiss the process by which experienced taxonomists "sense" evolutionary patterns. Despite the lack of explicitly stated analysis the taxonomist in question was undoubtedly mulling over some of the steps which have been clarified more recently to form the logical basis for the three recent schools: looking for pathways that minimize the number of evolutionary steps (parsimony methods), identifying groups of species which share derived character states (cladism), and choosing to ignore characters which show much homoplasy with others (compatibility analysis).

Parsimony

The first explicit methods for deducing the branching pattern of the evolutionary tree (known as the cladistic tree or cladogram) were based on the assumption of **parsimony,** that evolution followed the route involving the minimum amount of evolutionary change. If this evolutionary change is measured as length of lines on a tree, amount of change for quantitative characters and number of character state changes for qualitative characters, then the task is one of finding the branching tree of minimum total length. Such a tree will be an **undirected tree** (fig. 10.12) in graph theory as explained in chapter 5. Assigning evolutionary direction to the changes or locating the starting point of the trees are additional problems to be solved before a cladogram or **directed tree** (fig. 10.12) of the form desired by taxonomists can be produced.

Locating and recognizing the shortest directed tree presents an exceptionally difficult mathematical and computational problem and requires biological assumptions in addition to the assumption of parsimony. Cavalli-

Figure 10.11 A scheme for depicting present-day groupings of taxa as a cross-section through a phylogenetic tree. *(From Dahlgren 1980. Fig. 1).*

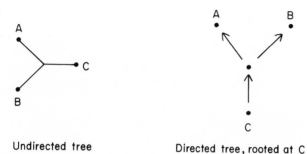

Undirected tree   Directed tree, rooted at C

Figure 10.12 An undirected tree ABC with no direction or starting point and a directed tree ABC.

Sforza and Edwards (1967) calculated that the number of possible trees from which the answer must be selected for only ten items is a staggering 34,459,425. Examining them all is beyond our powers of computation unless we can devote a dedicated and exceptionally reliable computer, such as a laboratory microcomputer, to run on the problem uninterruptedly for several weeks.

Methods which avoid examining all possibilities include the monothetic method (Camin and Sokal 1965) and the weighted invariant step strategy, "WISS" (Farris, Kluge, and Eckardt 1970). These give approximately minimal length directed trees but assume that the user can know the direction of character value or character state evolution for each character, and that each character follows an unbranched route. The assumptions are realized by requiring the user to code character values or character states upward from a primitive code of zero. Estabrook (1968) showed theoretically what was also being found empirically using the monothetic and WISS methods: that there are unfortunately a large number of equally short shortest directed trees for many sets of data.

To deduce the shortest undirected tree is a less severe problem (a mere 2,027,025 possibilities for 10 items) and does not require the user to make prior assumptions about the direction or route of evolution in particular characters. Consequently the **Wagner network** method (Wagner 1963; Farris 1970) was, for a time, the most widely applied explicit method. It assumes a geometric model in which points representing items are separated by distances recorded using, at least for quantitative data, the Manhattan distance already described in chapters 5 and 7. It will be seen from fig. 10.13 that this distance is equal to the number or amount of evolutionary changes needed for one item to evolve from the other.

The Wagner network is produced by finding the Steiner Minimum Spanning Tree, SMST, from the table of Manhattan distances between

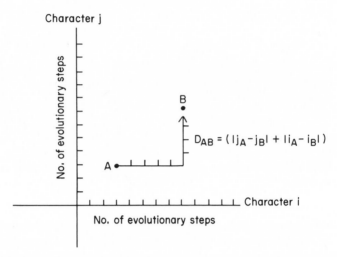

Figure 10.13 The Manhattan measure of evolutionary distances between two objects on the basis of two characters.

the items. This differs from the minimum spanning tree, MST, in two respects. MST is the shortest route drawn from item to item in the model, joining them all together in a connected tree without loops. The SMST is a similar tree, but made even shorter because some segments of the tree are allowed to be not between items, but between items and intermediate points called Steiner points fig. 10.14a. Introducing Steiner points in certain positions shortens the overall length of the tree and either introduces extra bifurcations or changes their position, as can be seen in the two-dimensional example in fig. 10.14b. The Steiner points may be thought of as hypothetical ancestors. Because the distances are Manhattan distances, we should get a more accurate visual impression of the tree's length if we redraw it as in fig. 10.14(c and d).

The resulting tree, or Wagner network, is undirected. To provide evolutionary direction or a starting point for the tree from biological considerations is often difficult and speculative. For instance we might speculate that the branch tips of the tree represent recently derived forms. In our small example this vastly reduces the number of possibilities to the four depicted in fig. 10.15, but even these look very different. Other considerations might be to examine B, C, X and D, for supposedly primitive character states, and for similarities to fossils or to members of other taxa.

The **Wagner tree** method differs from the Wagner network method in that a directed tree is produced. A character polarity must be given for each character and an ancestor specified for use as the root of the tree. Crisci

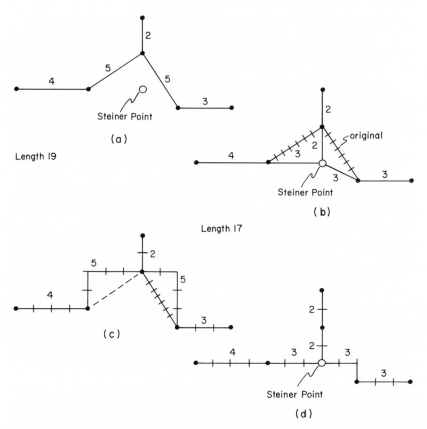

Figure 10.14 The use of Steiner points in producing the shortest minimum spanning tree. Note that the length of the two forms of the MST in (a) and (c) is 19 and that this is reduced to 17 in the SMST in (b) and (d).

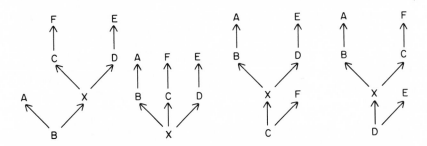

Figure 10.15 The effect of using different starting points when producing evolutionary trees from a Wagner network.

(1980) uses this technique to study possible evolutionary trees for 23 genera in the subtribe Nassauviinae in the Compositae. He uses four criteria for polarizing the characters. He interprets a character state as primitive if it is frequent within the group and widespread outside it, or if it covaries with other primitive character states. He accepts it as derived if it describes an adaptive specialization or if it is ecologically specialized and geographically restricted. After tabulating polarities based on the criteria for 27 characters, he constructs the tree with the minimum number of evolutionary changes, starting on different occasions from three different ancestral forms: a hypothetical ancestor combining all of the primitive states; *Dolichlasium,* the extant genus with the largest number of primitive states; and *Trixis,* a genus considered ancestral by previous workers. The three resulting trees shown in fig. 10.16 are markedly different in some respects (contrast the relative positions of *Leucheria* and *Polyachyrus*) and similar in others (such as the close linking of *Proustia, Lophopappus, Acourtia,* and *Burkartia* in all three trees).

       We believe that as methods for obtaining an authoritative reconstruction all of these are doomed by the assumption of parsimony. The evolutionary process, at least on the scale of microevolution, is one of opportunism; less is known about macroevolution, but an assumption that it twists and turns in unparsimonious fashion is more consistent with our present knowledge (Lowry, Lee, and Hebant 1980). Would not parsimony require functional whales to have evolved directly from fish, or functional cacti from the spurges? How could the first evolutionary steps possibly be part of an optimized path for completion millions of years later? Nobody, in fact, is making such biological claims for parsimony: its use is based more on the philosophical argument of "Occam's razor": the simplest explanation is usually the most plausible.

       The classification properties of these methods have not been fully explored. All three methods have mathematical relationships to phenetic methods (Wagner networks with single-linkage, the monothetic method with monothetic divisive cluster analysis, and WISS with median cluster analysis), but are sufficiently distinct to warrant further investigation.

Cladism

Comparisons of phylogenetic methods would be incomplete without consideration of techniques originated by Hennig (1950, 1966) and now popular as **cladism.** They are included despite being only partially explicit and involving manipulations which do not normally warrant computer assistance. In a sense the methods are not new: they present a precise exposition of threads that were evident in the sensing of the traditional school. Many English-speaking people have been slow to understand the methods: they were not published in English until 1966, and many concepts were given long, new names. Readers may prefer to read the simpler expositions of Gaffney (1979) and Humphries and Richardson (1980) before going to the originals.

The first step in a Hennigian analysis is to select a subset of the characters which are **synapomorphous.** The selection is applied to characters for which there are intuitive grounds for believing that one character state actually did evolve only once into a second, and in which the second is rare in the sense of being thought absent outside the group under study. Such characters must have a clear homology and a clear direction of change. A difficulty is that a number of arguments used to determine their polarity are circumstantial and not provable. A character that qualifies is thought of as a binary character **a** with a relatively primitive (**plesiomorphous**) state a_1 and a relatively derived (**apomorphous**) state a_2. It is the occurrence of two or more items that share the derived state a_2 that is known as a **synapomorphy** (a sharing of a derived state, ''syn-apomorphy''). So this step is finished by searching among the possible characters for characters with apomorphic states shared over several taxa.

The second step is to search through the synapomorphies in the hope of finding a nested sequence which we can pictorialize either as Venn diagrams or as cladograms.

In the Venn diagram of fig. 10.17a we think of each synapomorphous character as partitioning the items into two sets, the derived or apomorphous subset with derived character states shown within the circle, and the primitive or plesiomorphous subset outside the circle. The nested sequence we are seeking is simply a sequence of synapomorphous characters which have their derived subsets nested within each other as shown in fig. 10.17b. Alternatively we can draw the character distributions on a cladogram, but by adjusting the structure of the cladogram to fit the character distributions. Hennig (1966) calls this a ''scheme of argumentation.'' Figs. 10.17c,d give the equivalent cladograms for the data in fig. 10.17a,b. The boundary to each subset of items is now thought of as a bifurcation at which the apomorphic items acquired the apomorphic state and the plesiomorphic did not.

The difficulty that arises is that it may not be possible to find synapomorphies that nest. The synapomorphies may not nest, but instead conflict, as in the Venn diagram and cladograms of fig. 10.18 for synapomorphies **d** and **e.** Such conflicts can be resolved, but before we consider them we should first examine the analogy with evolutionary events. The cladogram of fig. 10.18(b) implies either that apomorphy e_2 has arisen inde-

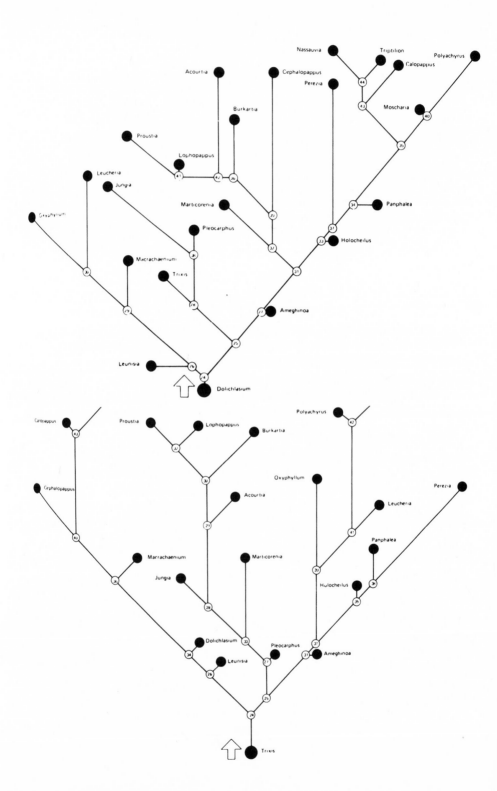

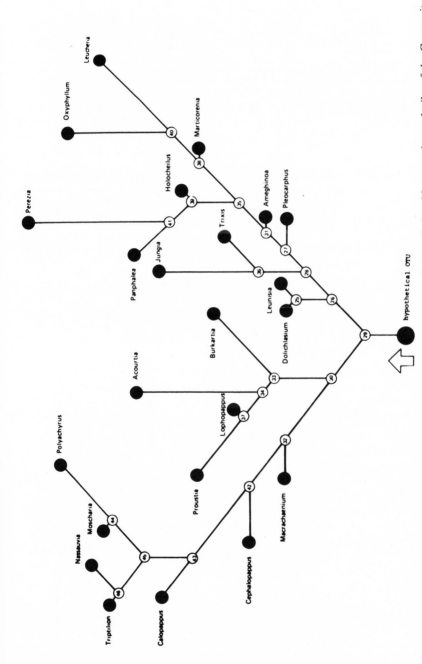

Figure 10.16 Three possible evolutionary trees derived by the Wagner tree method applied to 23 genera in a subtribe of the Compositae. Each tree is based on a different ancestral form. *(From Crisci 1980. Fig. 2, 3 and 1 respectively).*

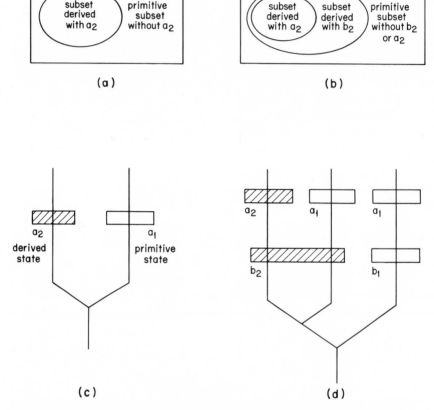

Figure 10.17 Venn diagrams (a and b) and corresponding cladograms (c and d) illustrate the derived (apomorphous) and primitive (plesiomorphous) subsets of items, and the nesting of apomorphous states sought in a Hennigian analysis.

pendently in items **J** and **K** (i.e., a parallelism) or that e_2 has reverted to e_1 in item **I** (i.e., a reversal). Conversely if fig. 10.18(c) is correct, then the parallelism or reversal will have been in character **d.** Unless we can resolve the conflict we cannot choose between the two configurations, and all four of the above evolutionary events are possibilities. (Using these conflicts to demonstrate parallelism or reversals is the basis for compatibility analysis described in the next section.)

One method of resolving the conflict is to use the majority vote. If, for instance, one of the synapomorphous characters is reinforced by the presence or the discovery of another synapomorphy of identical distribution, such as **f** in fig. 10.19, then these two are assumed to be correct and the

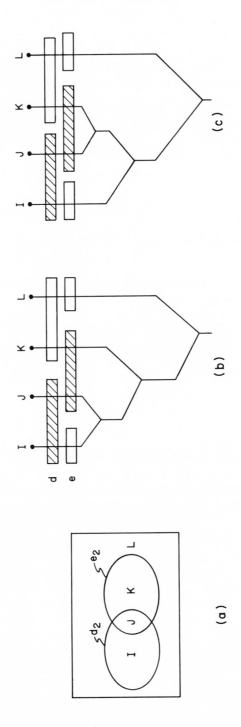

Figure 10.18 An example of conflicting synapomorphies—d_2 and e_2 in (a). If **I**, **J**, **K**, and **L** are items in the sets shown in (a), then there results an ambiguity as to whether cladogram (b) or (c) might be correct.

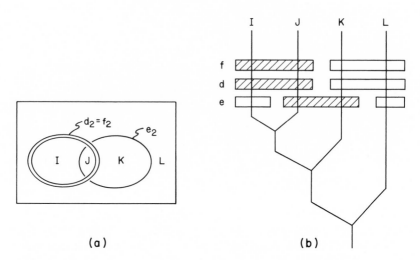

(a)   (b)

Figure 10.19 Resolving a conflict by majority vote. Synapomorphies d_2 and f_2 are found in the same items, but are in conflict with the single synapomorphy e_2.

once conflicting character to have a homoplasy. Cladists base this action on what they call the assumption of parsimony, but using the phrase in a different sense from that used in the earlier section of this chapter. Here it is assumed that, of alternative routes, evolution followed the one that required the fewest parallelisms and reversals, or that evidence supported by the most characters is most persuasive.

Another method of dealing with conflicts is to reexamine the grounds on which the conflicting characters were defined and on which they were selected as being synapomorphous. Dubious character-state delimitation might suggest improvements, thus changing the data and possibly the conflicts contained in the data. Where the user has more confidence in one synapomorphy than another, the other might be repolarized—that is, a fresh attempt be made at determining its polarity. It might also be polarized by reciprocal illumination—that is, arranging its polarity to agree with that of other characters, or it might be deleted from the study on the assumption that it was involved in homoplasy and was not therefore a uniquely derived synapomorphy.

The resulting cladogram, such as that in fig. 10.20 may be used by some biologists as a hypothesis about the evolutionary route for further testing and by others as a basis for classification. One property which affects its interpretation in both contexts is that it is monothetic. All the putative lineages or classes have at least one diagnostic or marker-derived character state. In evolutionary terms this implies that there exist at least some character states which throughout time, and in a multiplicity of organisms, have

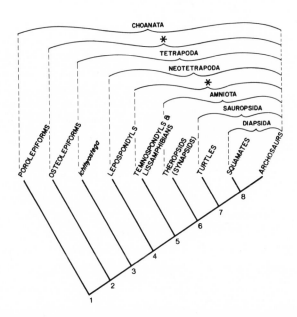

Figure 10.20 A classification (at the top) based on a cladogram (at the bottom) where acquired synapomorphies are labled 1–8. Note the large number of classes needed including two, marked with astersiks, that are new and must be named. *(From Gaffney 1979. Fig. 1, p. 90. Columbia University Press, by permission.)*

never reverted to an earlier character state. Think, for instance, of the diagnostic characters of mammals. At least one among the qualities of warm-bloodedness, hairiness, or the presence of mammary glands, is to be found in all present mammals. However, to satisfy this monothetic assumption it will have to remain true for future descendants. But in the comparable case of angiosperms, one of vessels, enclosed ovules, or double fertilization would have to be found in all present angiosperms and future descendants: already false. Either this means that vesselless angiosperms are ancestors of and outside the angiosperms, or there are reversals in which case presence of vessels ceases to be a synapomorphy.

Another property commonly found in cladistic results is for a single line of descent to have a separate node for nearly every unit analyzed as in fig. 10.20. If each node is used to form another rank in the hierarchy, as suggested in fig. 10.20, then the resulting classification has many ranks (often $N - 1$ for N units). Cracraft (1983) belittles this difficulty, claiming that the extra ranks are valuable in storing information on extra detail, and that some recent variants of cladism allow trichotomies, etc., as well as dichotomies.

We promised earlier to comment on each cladistic school in terms both of efficacy at inferring the true cladistic tree and, quite separately, of its usefulness as a classification technique. The cladist school, however, has developed a schism essentially along these lines—into those who use cladism in the context of phylogenetic studies, and those who use it as a pattern analysis prior to classification. To glimpse the full philosophical elaboration and intensity of these views see the extraordinary opposing papers by Hill and Crane (1982) and Charig (1982) in *Problems of Phylogenetic Reconstruction*.

The **classical cladists** (or Hennigian or evolutionary cladists) persist in interpreting cladograms in terms of possible phylogenies, the placing and sequences of fossils, the adaptive nature of synapomorphies and of biogeographic perspectives. The precision in cladist methodology is an enormous advance over narrative methods, but even so the method has, rightly, been criticized (Halstead 1982) for the apparent precision and respectability that a cladogram confers on what is really a speculation. Selecting synapomorphies, inferring polarity, and resolving conflicts are subjective, unreliable exercises, as Halstead (1982) demonstrates. We would also criticize the monothetic evolutionary model held by Hennig, although this has puzzlingly been dropped by some recent cladists. So, for instance, the presence of various chlorophylls might be proposed as a synapomorphy limiting several groups of green plants, even though there are substantial exceptions such as the lack of the chlorophylls in the parasitic angiosperm family Orobanchaceae.

By contrast the **pattern cladists** (or transformed cladists) offer cladism as the most effective hierarchical method of describing the variation pattern found among organisms. Synapomorphy is rephrased either as statements of homology (Patterson 1982), or as shared, less-generalized character states; and a cladogram is seen as representing the natural pattern or the classification, but not the phylogeny. They admit the uncertain tie between cladistic results and the true phylogeny. This view has much to commend it and, as we shall explain below, pattern cladism is increasingly being compared with phenetics in terms of precise information-preserving classification structures.

It should be recalled that the cluster analysis and ordination techniques often used in phenetic analysis actually place together pairs of items with the greatest overall similarity. However a group of items constructed in this way does not necessarily have many, or indeed any, character states common to all its members. An extreme case would be if items **A, B,** and **C,** scored for 100 characters, had a similarity **A** with **B** of 0.5, **B** with **C** of 0.5. It is possible for these to form a cluster (using single-linkage) at threshold similarity 0.5 and yet for there to be not one character state common to **A, B** and **C.** This is because although **A** and **B,** and **B** and **C** may be equally similar, the similarity may be based on different characters.

In practice of course this effect is usually less extreme—it means that group-ings formed by phenetic analysis are polythetic, defined on a combination of several characters where this combination is found in most, but often not all, of the members. In contrast, cladistic groupings are monothetic and so al-ways have at least one, and often many, character states (the synapomor-phies) found in all members. According to how one measures information content, such structures are claimed to outperform phenetic methods. They suffer, though, from the standard problem with monothetic groupings: the diagnostic character state may unite otherwise dissimilar items in what phe-neticists call artificial groupings. Placing the lungfishes with the tetrapods would be an example of this.

The question that has not been satisfactorily resolved is which of the two compromises, pattern cladism or phenetics, gives the better in-formation properties to the classification. The ideal would be a grouping in which a large number of diagnostic character states are present in all mem-bers. Such a grouping would of course be so clearcut that its classification would be beyond contention, and also both phenetic and cladistic methods would yield the same result. (Such groupings exist, but their delimitation is easy and therefore rarely discussed.) The problem is what to do when the pattern of variation is more complex. The phenetic methods group items on overall resemblance, relaxing the need to have diagnostic character states present throughout the group, while the cladistic method maintains at least minimal diagnostic character states, but relaxes the need for members to be similar on overall resemblance.

Compatibility

This approach depends on a logical method for detecting homoplasies (par-allelisms and reversals) suggested by Wilson (1965) and LeQuesne (1969), and much developed by Estabrook (1972, 1978, 1980) and LeQuesne (1982).

Let us examine the combinations that might occur for two bi-nary characters, a, with states a_1 and a_2, and b with states b_1 and b_2. Four character-state combinations are possible $(a_1b_1, a_2b_1, a_1b_2, a_2b_2)$. LeQuesne (1969) pointed out that if any two or three of these combinations occur in nature then we can construct one or more evolutionary trees apparently free from parallelism and reversal. But in contrast, if all four combinations are found, then all of the trees that might be postulated must contain at least one parallelism (that is, a change such as a_1 to a_2 occuring twice) or one reversal (that is, a change such as a_1 - a_2 occurring in both directions).

Readers should satisfy themselves of this—either by continuing the diagram above until all possible trees have been enumerated, by referring back to the conflict condition discussed under cladism earlier in this chapter, or by the following proof. There are only two evolutionary changes that may occur, that between a_1 and a_2 and that between b_1 and b_2. If four combinations are found, three evolutionary steps are needed to join them in a tree. As only two evolutionary changes are possible, the third evolutionary step must be a repeat or a reversal of one of the other two.

Notice that although for four combinations there must have been a parallelism or reversal, we cannot know whether it was in character **a** or **b**. For instance in fig. 10.21 the various possibilities cannot be distinguished. In the left hand example there is a reversal in **b** but in the next to left example there is a parallelism in **a**. Such a pair of binary characters is said to be **incompatible** because an evolutionary tree deduced from the one character $(a_1 - a_2)$ is logically incompatible with that deduced from the other $(b_1 - b_2)$. Pairs of characters which do not necessarily have a parallelism or reversal are said to be compatible (Estabrook 1972) or uniquely derived (LeQuesne 1969).

Estabrook has worked extensively on developing the detection of incompatibility for multistate characters as well as the detection not only of pairs but also of whole sets of compatible characters known as **cliques.**

For binary data the Estabrook and Meacham (1979) method detects and lists all of the cliques of compatible characters. Figure 10.22 from Estabrook (1980), shows an exceptional example in which the method has

Undirected trees for 3 combinations

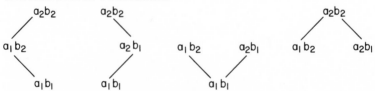

Undirected trees for 4 combinations

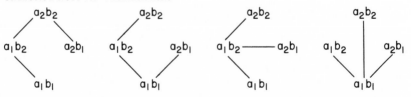

Figure 10.21 Possible combinations into undirected trees of two binary characters **a** and **b**. Note that in the trees for 3 combinations there are no parallelisms or reversals, while in those for 4 combinations each will have at least one parallelism or reversal.

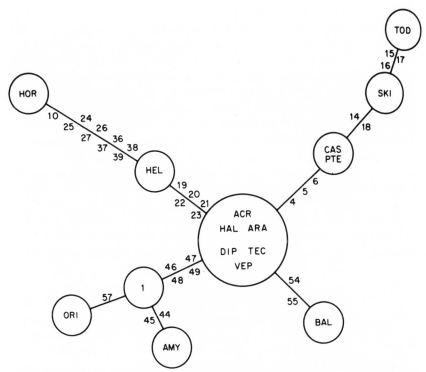

Figure 10.22 An undirected tree formed by the largest clique of 14 compatible characters, with terminal segments added for a further 17 compatible characters found in 1 or all but 1 of the taxa. (Taxa are labelled by three-letter codes, characters by numerals.) *(From Estabrook 1980. Fig. 3).*

discovered a clique of 14 binary characters (actually the presence or absence of 14 chemical skeleta) which, among 15 taxa, are present in anything from 2 to 14 of them. After placing the taxa in circles, if they have the same combination for the 14 characters, it is possible to join the circles by links which represent changes in one or more of these 14 characters, as in figure 10.22. A further 17 characters in this exceptional data set were present in either 1 or 14 of the 15 taxa. Such characters are compatible with all other characters and merely add terminal segments (branch tips) to the tree, as in fig. 10.15. So for example taxon ORI is distinguished from all others by the presence or absence of character 57, and consequently is placed on a branch tip. It should be noted that the effect of the compatibility analysis in this example has been to eliminate the 33 characters which are incompatible with the maximal set—that is, to eliminate what are thought to be the characters responsible for reticulation. The tree produced from this now ultrametric data set is the dual case—a Steiner minimum-spanning-tree which is both a single-linkage classification and a most parsimonious Wagner network.

So we see from this brief survey that while great progress is being made in trying to solve the problem of phylogenetic reconstruction, no method as yet can provide an authoritative solution; each is flawed by the biological assumptions needed: parsimony in its various forms, a knowledge of character state phylogeny, a knowledge of uniquely derived characters, or a knowledge of polarity of character state change.

The incompatibility analysis is of particular interest. It analyzes homoplasy directly whereas the other methods would work well but for the complications that homoplasy causes. The Hennigian analysis is of particular interest because it alone derives all present-day forms from ancestors, and because even if one disregards its utility in tracing evolution, it can be used to produce a classification with interesting properties.

Suggested Readings

General

Felsenstein, J. ed. 1983. *Numerical Taxonomy.* Berlin: Springer Verlag.
Hennig W. 1966. *Phylogenetic Systematics.* Urbana. University of Illinois Press.
Joysey, K.A. and A.E. Friday, eds. 1982. *Problems of Phylogenetic Reconstruction.* London: Academic Press.
Nelson G. and N. Platnick. 1981. *Systematics and Biogeography.* New York: Columbia University Press.
Wiley, E.O. 1981. *Phylogenetics: The Theory and Practice of Phylogenetic Systematics.* New York and Chichester: Wiley Interscience.

Part IV

Computer-Assisted Database Management

Chapter 11

Database Management

Another essential activity for taxonomists is data management. Means for data capture, data storage, and data retrieval and the role of computers in accomplishing these functions will be discussed in this final part of the book. We will begin with a very brief introduction to computers themselves—the so-called hardware. It is difficult to talk about computers in generic terms so that in places specific computers and programs are named. Our purpose is only to give concrete examples, not to provide a complete list or to make recommendations. In any case it is generally true that taxonomists' needs are not specific and unique with respect to hardware; many different kinds of machines can be used successfully. Further, truly phenomenal changes are taking place so rapidly that it would be foolish for us to compare hardware except in the most general terms. Significant trends in hardware are occurring: costs are down; speed of operation is up; and memory capacity is larger and cheaper. All of these trends have enormous impact on the feasibility of computer applications in taxonomy as well as in other fields.

Computing Facilities

Computers can be classified roughly according to size. Mainframes were once housed in their own suites of air conditioned rooms and attended by resident computer experts. They could be used offline by depositing a deck of cards with the operator and waiting for the response to be printed out on a line printer, a procedure known as batch processing. After initial input from cards large data decks were stored on magnetic tapes or disc packs. When this was done batch processing, even of large data sets, could be done online using a terminal. Today the majority of mainframe computer activities are carried out from a terminal online and interactively; that is, with the work being performed immediately as the instructions or data are entered (online) and with programs that chat back to the user (interactively) rejecting or accepting commands, prompting further commands or data and so on.

Computing needs of many taxonomic projects are still met in this way by large off-site mainframes. An example is PRECIS, a large database at the National Herbarium, Pretoria, South Africa. There the computer used is a Burroughs 7800 at the Department of Agriculture (Gibbs, Russell, and Gonsalves 1984). In the Vicieae Database Project at Southampton the main databases are run on the University's ICL 2960 mainframe, but are accessed online from either terminals or micros in the taxonomy laboratory.

The advent of the chip and the miniaturization that followed led in the 1970s to a new generation of computers called minis (minicomputers), a term which refers to their physical size and cost rather than to their performance. They could be used as dedicated computers for large projects, and are normally used from video terminals. The Digital Vax 11/780 used for Flora Veracruz at Jalapa, Mexico and the DEC PDP 11/34A at the ESFEDS project in Reading, U.K. are both examples of minicomputers devoted to major floristic projects, which we shall describe in chapter 12.

Today's era is that of the micro (microcomputer)—even smaller, reliable, inexpensive machines that range in size from home computers (e.g., 16K or 28K Sinclair, Atari, or BBC-B home computers), through the medium-sized business or personal computers (e.g., 64K, 128K, or 256K RAM machines such as Kaypro, Apple, Sirius, or IBM PC with operating systems such as CP/M, MS-DOS or PC-DOS), to the larger machines (e.g., 256K or 1028K machines such as Apricot and IBM XT, with fixed Winchester discs, also called hard discs, and operating systems such as CP/M 86, MSDOS or PCDOS or UNIX). For the larger machines the processing speed "is roughly equivalent to that of the minicomputers which came onto the market at the beginning of the nineteen seventies" (Freeston 1984).

Small and medium machines may often be dedicated to a single person or research group or to a single task, and are extensively used for word processing. They also may be used instead of terminals for minis or mainframes. They are normally operated using a keyboard and video monitor, and with one or two floppy disc drives and a printer. Large minis, and many of the medium-sized minis, are operated with an additional fixed disc of from 5 to 50 megabytes. They may be used to support quite extensive databases, and in many examples are used as multi-user machines with other machines attached. To take advantage of the large amount of ready-made software available, purchasers should consider either machines that are already very popular or machines that employ the widely used operating systems.

Micros have revolutionized computing in many scientific and business establishments. The combination of user-friendly software and reliable hardware means that most machines are now used directly by scientists and businessmen without professional assistance. And in large institutions they have the advantage of leaving the user complete freedom to start and stop as and when he likes. Using a micro instead of a terminal for accessing a mainframe has advantages too. With a terminal emulator the micro can be used just as a terminal to the mainframe, but if and when desired data can be transferred to or from the micro. Thus, with a micro, one might create and edit a data file on one's own time, and then connect up and enter the file for use on the mainframe. Or, as in the Vicieae Database Project, copy the results of a mainframe database inquiry onto the micro for word-processing and tidying and editing before publication or dispatch to users. As much of the commercially available software is marketed very widely, an additional bonus is that databases prepared with a particular database management program, like "dBase II" or "Knowledgeman," may be easily transferred from one machine to another within or between institutions.

Databases

A **database** is simply the set of data pertaining to a particular purpose or project organized for multiple and repeated use and intended to serve many purposes and many people. Databases typically contain tabular data referred to as **flat files** or simply **files.** There may be one or more tables of data in a database. If there are multiple files, they are linked together by some relationship. These may be hierarchical or network kinds of links among the files, or the files may contain common elements which act as linkers.

Within the file there are **records,** which consist of the data elements pertaining to a single item or case. A record, in our examples (chapter 2), occupies a column in a character by taxon table; in other systems a record is equivalent to a row in the same kind of table. But it really makes no difference which convention is used. Each record is divided into elements known as **fields.** Each field contains a descriptor or character state appropriate to a particular item.

Whether or not a computer is to be used, building a database requires careful planning before the collection and recording of data. It is not sensible to begin without a firm, well-defined objective in mind. Too many databases arise from an ill-defined expectation that any and all data will be useful somewhere, sometime. While it is reasonable to hope for a degree of serendipitous exploitation of databases, "pack-ratting" alone should hardly constitute the major focus for data collection.

Despite the Utopian expectations of some a decade ago, the "computer does not change the basic human variables in the taxonomic equation" (Shetler 1975). Neither does it make it possible to attain some ultimate goal of cataloguing and classifying all living forms. As we have seen, what the computer can do best is to manipulate large sets of data so as to provide new syntheses and information. Thus, it seems reasonable that as much as possible of the considerable amount of work that goes into the collation of data be expended in producing as general a database as is practical. Such a database is the kind that can be analyzed in several ways and can deal with more than the single problem which motivated it. The collection and organization of data is difficult and labor intensive. We should use it wisely.

One of the most common and difficult problems faced at the point of analysis is missing data entries. It is often good strategy, when collecting and recording the information for a general purpose data bank, to devise a format to help ensure complete data gathering. A form that lists all of the variables in such a way that their values can be filled in or checked off at the time of recording each item is invaluable. Fig. 11.1 gives two examples of recording forms used for data collection. The examples from PLANTAX (Sweet and Poppleton 1977) and the Colombia National Herbarium are clearly designed to facilitate transfer of the data to the standard 80 columns available for data in many computer programs.

Preset collection formats like these are useful not only for reasons of computer input. They also help to overcome the human tendency to record only what is new and different after one has examined a large number of specimens. In addition to having complete data, it is important to have the data in standard form so that the items are strictly comparable. The discipline must exist to record all variables for each item in a predetermined form. It may be a tedious process, but without this conceptually simple step the more sophisticated outcomes can be doomed from the start. The lack of such data collecting standards has been the stumbling block for many plans to combine data from different laboratories and/or museums.

When building large, general-purpose databases, especially when they contain data on widely varying items such as all the vascular plants, some variables are bound to be logically inapplicable. Such states, recorded as inapplicable, need to be recorded in such a way as to be distinguishable from other categories of the no response or zero type, which can indicate that values are unknown or are actually equal to zero.

The data collected and recorded in some primary fashion must be transferred to the computer. As a general rule the shortest path from the source of the data to the computer reduces errors and is most efficient. A computer terminal or microcomputer is often well suited to data recording, especially when there are associated editing or word processing facilities.

There are some formal aspects of the structuring of data for processing with a computer that we must consider when planning the database. One is in regard to **coding.** Until recently, space in computer memory was a major cost consideration and had to be used efficiently. One response was to cut down data input by devising codes for either descriptor names or states or both. There are many programs still in use—for example, the one represented by fig. 11.1b—that were originally designed for the 80-column punched card and required coding. However, the present availability of multimegabyte hard discs goes a long way toward obviating the requirement for coding to save space in computer memory.

Data that serves as input to the computer must be presented in proper order. The layout of the data, and particularly of the symbols, spaces, and separators needed to communicate this layout to the data-management program, is called the input **format.** Appropriate format is always precisely specified in the instructions for the program that is used. In general, data are entered into fields in one of two types of format, using either preset field lengths or designated characters as separators (as commas are used in the TAXIR example below). Of course, either type of input format requires that the order of descriptors be rigorously consistent, since order, in both cases, indicates the correct association between descriptor and data field.

Computerized Database Management

Taxonomic databases are not intrinsically dependent on computers; they have been used for centuries. However, when computers began to be available for general use two decades ago, taxonomists realized that computers could be of enormous benefit in saving time and human effort during the processing of taxonomic data. Taxonomists began to experiment with using computers and the ancillary devices associated with electronic data processing in the '60s, and by 1970 "An index of EDP-IR projects in systematics" (Crovello and MacDonald 1970) was published that contained summaries of 40 such projects.

Early on the interest in applying electronic data processing methods to taxonomic data led to investigations of general information storage and retrieval systems, which were then known as EDP-IR systems or

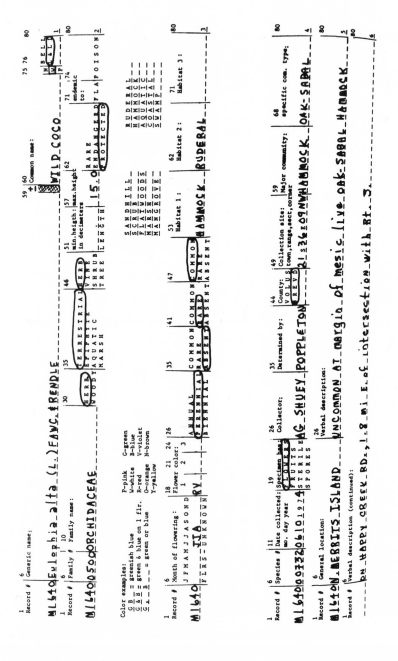

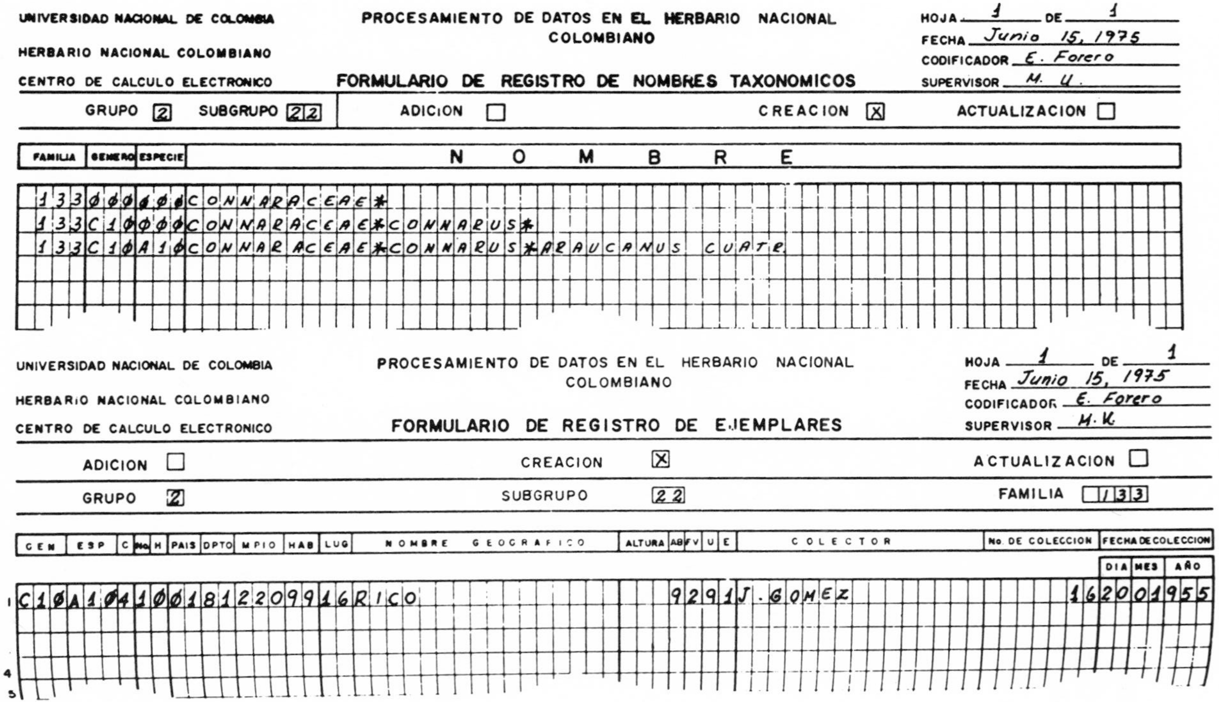
UNIVERSIDAD NACIONAL DE COLOMBIA
HERBARIO NACIONAL COLOMBIANO
CENTRO DE CALCULO ELECTRONICO

PROCESAMIENTO DE DATOS EN EL HERBARIO NACIONAL COLOMBIANO

FORMULARIO DE REGISTRO DE NOMBRES TAXONOMICOS

HOJA 1 DE 1
FECHA Junio 15, 1975
CODIFICADOR E. Forero
SUPERVISOR M. U.

GRUPO [2] SUBGRUPO [22] ADICION [ ] CREACION [X] ACTUALIZACION [ ]

FAMILIA	GENERO	ESPECIE	NOMBRE
133	ØØ	ØØ	CONNARACEAE*
133	C1	ØØ	CONNARACEAE*CONNARUS*
133	C1	A1	CONNARACEAE*CONNARUS*ARAUCANUS CUATR.

UNIVERSIDAD NACIONAL DE COLOMBIA
HERBARIO NACIONAL COLOMBIANO
CENTRO DE CALCULO ELECTRONICO

PROCESAMIENTO DE DATOS EN EL HERBARIO NACIONAL COLOMBIANO

FORMULARIO DE REGISTRO DE EJEMPLARES

HOJA 1 DE 1
FECHA Junio 15, 1975
CODIFICADOR E. Forero
SUPERVISOR M. K.

ADICION [ ] CREACION [X] ACTUALIZACION [ ]

GRUPO [2] SUBGRUPO [22] FAMILIA [133]

GEN	ESP	C	No	H	PAIS	DPTO	MPIO	HAB	LUG	NOMBRE GEOGRAFICO	ALTURA	AB	FV	U	E	COLECTOR	No. DE COLECCION	FECHA DE COLECCION DIA	MES	AÑO
C1	ØA	1	Ø4	1	ØØ	18	122	Ø9	9	16	RICO		9	2	9	1	J. GOMEZ	16	2Ø	Ø1	955

Figure 11.1 Two examples of forms for data collection. (a) Form which serves as a check list for descriptive data and also includes taxonomic and collection data. *(From Sweet and Poppleton 1977. Fig. 1).* (b) The upper form holds taxonomic names and the lower one specimen data in coded form. *(From Forero and Pereira 1976:86).*

simply IR systems. Such systems were seen to be made up of two subsystems—one to store the data and a second to retrieve it. These two functions are still the basis of information management systems today, although terms like IR are now out of vogue and we talk about searches instead. As is summarized in fig. 11.2, the data storage function consists of creating a file—that is, entering the data and then editing or updating that file. Using the data involves the retrieval function, which is made up of querying and output components.

In the rest of this section on database management systems we shall consider these two essential functions—storage and retrieval—in relation to a number of database management programs that are available for computers. There are, of course, a very large and ever-increasing number of such programs, and it would be easy for us—and therefore our readers—to get bogged down in the operating details of particular programs. To avoid this we have chosen to use a program called TAXIR, for TAXonomic Information Retrieval (Estabrook and Brill 1969), and later EXIR (Abbott 1975) as a running example of a flat file management program. As a practical system for database management TAXIR has been largely replaced by newer programs. However, it was one of the first computerized information systems; it was written expressly for use with taxonomic data, and it is very simple and straightforward in its structure. This simplicity is our principal reason for using it here. As a pedagogical device TAXIR is a particularly clear and instructive example of the processes that go on in all data file management.

After discussing data storage and retrieval for single files, we shall look at management for combinations of files using other programs as examples. All of these procedures involve the use of computers, which, even in the 1980s, may not be an everyday experience for everyone. So it is important to note that the user need not know anything about computer programming, or about how the computers work, to be able to use computers for database management.

Data Storage

The first step in data storage is, obviously, to enter the data, which has been planned and collected, as discussed above, into the computer. Until the last few years the most common method for initial data input was by means of punched cards. These cards, produced by a keypunch machine and with sets of holes that code for alphanumeric characters, were submitted to the computer in the form of a deck for batch processing.

Remote input devices are the major routes into large mainframe

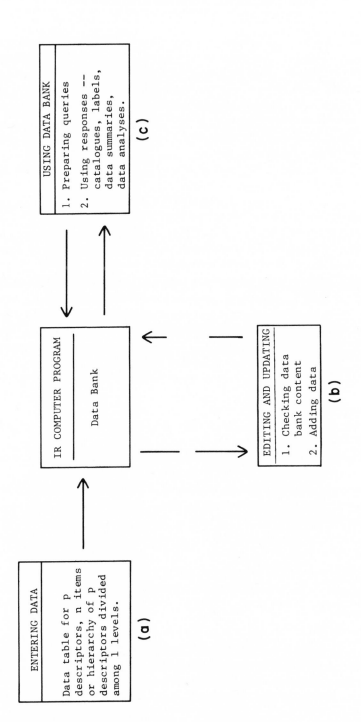

Figure 11.2 Flow chart of a data storage and retrieval system.

computers today. A terminal or microcomputer (personal computer), connected to the computer either by direct wiring or by a modem and telephone lines, may be used. In either case, the user simply types the data in at the terminal or micro, and it is transferred immediately as input to the mainframe or minicomputer. An alternative is off-line typing of data onto magnetic tapes or discs, then rapid on-line transfer to the central computer, thus reducing charges for central computer time and communications.

Most modern database management systems provide both customized display screens and screen editing. Customized display screens are formatted, pre-arranged layouts on the screen of a terminal or micro. They are arranged so that the screen looks like a business or administrative form, with places to fill in or display the data, and notes and prompts. In taxonomic examples the notes often provide reminders of the states of each character. At the time of data entry the user has to fill in the boxes; that is, for instance, type in the name of a species in the appropriate box or a score on character states against the notes on each character. Later on the same screen can be used for examining data about a given species, or for editing the data. A screen editor provides facilities similar to those incorporated in word-processing programs—the user can move the cursor freely about the screen and make additions, substitutions or deletions where needed.

Once the initial data entry has been accomplished there are a number of media which accomplish off-line, mass storage. On mainframes magnetic tapes and disc packs are used for this purpose. On microcomputers, one can use floppy discs or even cassette tapes if relatively small amounts of data must be stored, or hard discs (Winchesters) for storing the larger data files. In all cases, the storage media also serve as input devices since database files, once created, must be reentered from time to time.

Before the actual process of data input can begin one must define the descriptors that will be used and delineate the items that will be the subjects of description. In fig. 11.3 there is a small, sample data set that we will use in our TAXIR example below. The reader will recognize the tabular data structure. The data in the internal cells of the table are the elements that are actually processed by the computer. This data, along with the row headings, which are the descriptor names, must be entered into the computer memory in the form that is required by the information management program, in this case TAXIR.

Data Input in TAXIR

In TAXIR the DEFINE DESCRIPTORS and DEFINE ITEMS commands control the data input process. The instruction DEFINE DESCRIPTORS first indicates the number of descriptors that will be entered. Then it lists the names of these descriptors and numbers them in the order

BINOMIAL	C. incana	C. incana	C. agatiflora	C. nitens	C. pilosa	C. nitens
ACCESSION NUMBER	1	2	3	4	5	6
ALTITUDE	200 m	850 m	2600 m	2100 m	2000 m	1500 m
HABITAT NOTES	Not recorded	Wasteland in village	Terraced garden	Road edge	Roadside near spring	
COUNTRY	Peru	Colombia	Colombia	Colombia	Colombia	Peru
LIFE FORM	herb	herb	shrub	herb	herb	herb

Figure 11.3 Sample table from a database for *Crotalaria*.

in which they will appear for each item. Since TAXIR can handle any kind of qualitative or quantitative data, a parameter must be included with each descriptor to indicate to the machine what kind of data to expect for that descriptor. Examples are ORDER as the parameter option for quantitative data and NAME for simple qualitative multistate (see figure 4.1) such as COUNTRY in the example. The program requires that each descriptor be listed, numbered, and followed by its type parameter. In our example the descriptors were entered in the following form:

LIFE FORM(1,NAME),BINOMIAL(2,NAME),
ACCESSION NUMBER (3,ORDER),ALTITUDE(4,ORDER),
HABITAT NOTES(5,NAME),COUNTRY(6,NAME)*

Note the use of the comma and asterisk. As is the case with most programs of this sort, the specified format and punctuation must be followed exactly, since it is the symbols and order that enable the computer to store data in the correct places in its memory.

Once the descriptors are defined, we are ready for data regarding the items to be entered into computer memory. An item consists of a series of descriptor state fields ordered so that each contains the data related to one of the descriptors previously defined. In our example in fig. 11.3 each column defines an item.

After the two types of data input—descriptors and items—are ready on the input device, a data file can be created in the computer by a few simple commands, which, like the data, can be punched onto cards or typed in at the terminal. The result of the DEFINE AND PRINT ITEMS command is an output such as in fig. 11.4, which is useful for verifying the input.

Editing and Updating in TAXIR

After input, one needs to verify that the data have been entered correctly. Further, the means for correcting errors, that is, **editing,** are needed. Special commands exist for the correction of the data entered for individual items. We shall give some examples below.

There must also be ways to **update** the data. This means having the capacity for adding new items and deleting old ones. There is further need for adding new descriptors and the data pertinent to them for items already in the database. This process can be seen as roughly equivalent to correction.

Errors may be corrected by a CORRECTION command. In our example, we could change the altitude for *C.pilosa* to 2100 m by one of the following correction statements:

CORRECTION (ALTITUDE, 2100) FOR THOSE WITH ACCESSION
NUMBER,5*
CORRECTION (ALTITUDE, 2100) FOR THOSE WITH BINOM-
IAL,C.pilosa*

Means for updating TAXIR include the capacity to add new items. This is a simple process of typing the new items as if they were the original ones. They are then submitted using the DEFINE ITEMS command again. The new items will now be included with the original ones and used in all subsequent manipulations of the database.

Items can be deleted by a command which will either eliminate

```
0023 DEFINE ITEMS FROM CARDS FREE*
0024 Herb,C.incana,1,200m.,Not recorded,Peru*

**ERROR TYPE  50   COLUMN 20    DESCRIPTOR NO.   4
0025 Herb,C.incana,1,200,Not recorded,Peru*
0026 Herb,C.incana,2,850,Wasteland in village,Colombia*
0027 Shrub,C.agatiflora,3,2600,Terraced garden,Colombia*
0028 Herb,C.nitens,4,2100,Road edge,Columbia*
0029 Herb,C.pilosa,5,2000,Roadside near spring,Colombia*
0030 Herb,C.nitans,6,1500,,Peru*
0031 END OF ITEMS*

 ENTER NEXT COMMAND
0032 WRITE DATA BANK CROT1*
```

Figure 11.4 Card image output resulting from item definition in TAXIR.

single items or special subsets of items that have some character state in common. Examples of such commands applied to our data set are:

DELETE ITEMS WITH COUNTRY,Peru*
DELETE ITEMS WITH ALTITUDE(FROM 200 TO 1000 IN m)*
DELETE ITEMS WITH (ACCESSION NUMBER,1)*

The DELETE statement is a powerful command in TAXIR and must be used with some care. If the second example had been written (FROM 200 to 10000) for instance, the whole set of data would have been erased from the memory!

The TAXIR commands for input and editing are cumbersome to use when compared with those of the newer database management programs. And the possibility that a single blundering command to delete could wipe out an entire database has now been eliminated by buffering and user-owner security systems! What has *not* changed in database management systems since the days of TAXIR (only 10 to 20 years ago at that) are the essential principles. The reader should realize that any modern system will operate on one or more data tables, that there will be a means for entering descriptor names and linking them to appropriate descriptors and descriptor states for each item. The means for this may be a formal formatting requirement, as in TAXIR, or it may be a prompt to enter data at a given place or given time on a custom-arranged screen. On modern systems, too, the means for editing the data in a file, particularly by using a screen editor, are likely to seem much less formidable than the command cards used in TAXIR. However, by whatever vehicle, editing files in a database management system will require, first, a descriptor state used as a selector to pick out the item to be corrected and, second, a means for specifying both descriptor and descriptor state pertaining to the data to be added or substituted.

Information Retrieval

Once the data have been stored, the problem is to retrieve those parts that are needed. Usually users do not want to see all the data in a file or even the whole list of data for a particular item. Instead, they want to select that portion of the information that is relevant to a question they want answered. The procedure for doing this is called **querying** the data. As the name implies, it is a process for asking questions of the computer. The output of the computer, in response to a query, can be a simple list or it can be a multi-columned catalogue of the requested descriptor states that apply to a selected subset of the items. The form of the question is strictly defined, so as to make it recognizable to the computer as an instruction. The reply is equally rigidly structured in TAXIR, although some of the more recent "user friendly" programs manipulate the formal output into a form that is nearer to a common language reply.

Retrieval in Flat Files

Searching a database can involve more than querying a single data table, as we shall see later in this section. However, in order to better understand basic information retrieval procedures, we shall begin with a description of the process as it works for a single file. Continuing with our TAXIR and *Crotalaria* example, we analyze the following sample query:

PRINT: ACCESSION NUMBER FOR ITEMS WITH COUNTRY,Peru*

and the response to it:

3 OUT OF 7 ITEMS HAVE THESE ATTRIBUTES (42.86 PER CENT).
  1
  6

We note that the query instructs the computer to pick out of all items in the database file those that are found in Peru, where Peru is one of the possible states of the descriptor COUNTRY. This illustrates what is a general principle of retrieval: a subset of the items is designated by using one or more of the descriptor states as a selection criterion.

The remainder of the instruction tells the computer to look up and then print the ACCESSION NUMBER for each of the items in the subset. This is again a general procedure: one or more attributes of those items in the subset are designated as the output list.

The query structure illustrated is in its most elementary form. It can be expanded. One possibility is to form the subset on a selection criterion that is not a single descriptor state but a combination of descriptor states from either one or several descriptors. These criterion descriptor states are joined by the Boolean operators **or, and,** or **not.** In the TAXIR implementation we are using (but not in all programs), the operator symbols are these words surrounded by periods, giving the character strings: .OR., .AND., and NOT. (Since the entire string represents the operator, the embedded periods/full stops are not confused with periods/full stops actually within the data.) Another means for variation on the elementary form of query is the HOW MANY phrase. It commands the computer to count rather than to list the items in a designated subset. The results are shown below each one for the following examples of queries:

HOW MANY ITEMS HAVE (COUNTRY,Peru).AND.(LIFE FORM,Herb)*
3 OUT OF   7 ITEMS HAVE THESE ATTRIBUTES (42.86 PER CENT).

HOW MANY ITEMS HAVE (BINOMIAL,C.nitens.OR.C.pilosa)*
3 OUT OF   7 ITEMS HAVE THESE ATTRIBUTES (42.86 PER CENT).

Note that the resulting subset of items for the first query includes those items which are herbs growing in Peru—that is, the states Peru and Herb apply simultaneously to the items. In the second query above each of the items designated for inclusion in the subset must belong to one *or* the other of the species indicated as the state of the descriptor BINOMIAL.

A further important expansion of the querying capacity occurs when one includes several variables in the output list. Fig. 11.5a shows the output for such a query in a case in which a list of variables is called for. In the example in the figure the printout was produced on a standard line-printer.

The reader may have noticed that our outputs up to now have had seven items indicated instead of the six that are shown in the data table. This is due to the error noted in the listing shown in fig. 11.4. It calls for a correction which was done using the DELETE ITEMS command shown in fig. 11.5b. Several responses to other queries of the corrected database are shown in fig. 11.5c. Here the EXIR output has been printed on a word processor instead of on a line-printer, so that some program commands and line numbers have been edited out.

In addition to these standard queries, there are several other forms. Catalogues can be printed directly from the database in a report format specified by the user. An auxiliary program (e.g., EXIRPOST in the Vicieae Database example in chapter 12) can be used to generate headings, calculate column widths, arrange pagination, and so on so that the results of the query are printed in report form.

If individual items are to be retrieved from computer memory, it is logically necessary that they have at least one descriptor, which has a unique state for each item, to act as an identifier. Some programs require item identification in a specified format and/or location in the input; such identifiers are intrinsic to the program in that they serve as tags for locating the items. The TAXIR program, like many modern ones, has no such requirement and regards an identifier as no different from any other descriptor.

As we have shown in the examples above, an item can be included in a query response on the basis of any of its descriptor states. It is not necessary to retrieve an item via a single route such as its identifier. This capacity is termed **multiple access.**

An information system can be thought of as an extended card file. It is physically possible to arrange index cards in only one order at a time. To do this sorting we use a single descriptor as the selector criterion, for example, ITEM or NAME or SPECIES. The cards can be rearranged on the basis of a different criterion like HABITAT or COLOR, but the process

```
ENTER NEXT COMMAND
0042 PRINT:BINOMIAL,ALTITUDE,LIFE-FORM,COUNTRY,ACCESSION NUMBER,HABITAT NOTES
0043 FOR ENTRIES WITH LIFE-FORM,Herb.OR.Shrub*

       7 OUT OF      7 ITEMS HAVE THESE ATTRIBUTES (100.00 PERCENT).

C.agatiflora
      2500 m.
            Shrub
                  Colombia
                        3
                        Terraced garden



C.incana
      200 m.
            Herb
                  Peru
                        1
                        Not recorded
      350 m.
            Herb
                  Colombia
                        2
                        Wasteland in village
      ***
            Herb
                  Peru
                        1
                        Not recorded
C.nitens
      1500 m.
            Herb
                  Peru
                        6
                        ***
      2100 m.
            Herb
                  Colombia
                        4
                        Road edge
C.pilosa
      2000 m.
            Herb
                  Colombia
                        5
                        Roadside near spring
```

Figure 11.5 Examples of query outputs in TAXIR.
(above) Output for query with list of variables.
(upper right, next page) Output for correction using DELETE ITEMS.
(lower right, next page) EXIR output using word processor.

```
  ENTER NEXT COMMAND
  0044 DELETE ITEMS WITH ALTITUDE,UNKNOWN*

     1 ITEM(S) ARE TO BE DELETED FROM THE DATA BANK.


  FORMER NO. OF ITEMS IN THE DATA BANK  =    7
  NO. OF ITEMS DELETED                  =    1
  CURRENT NO. OF ITEMS IN THE DATA BANK =    6

  ENTER NEXT COMMAND
  0045 PRINT:(BINOMIAL,ALTITUDE,LIFE-FORM,ACCESSION NUMBER,COUNTRY,HABITAT NOTES)
  0046 FOR ENTRIES WITH LIFE-FORM,Herb.OR.Shrub*

      6 OUT OF     6 ITEMS HAVE THESE ATTRIBUTES (100.00 PERCENT).

  C.agatiflora 2600 m. Shrub   3 Colombia Terraced garden
  C.incana      200 m. Herb    1 Peru     Not recorded
  C.incana      850 m. Herb    2 Colombia Wasteland in village
  C.nitens     1500 m. Herb    6 Peru     ***
  C.nitens     2100 m. Herb    4 Colombia Road edge
  C.pilosa     2000 m. Herb    5 Colombia Roadside near spring
```

1
6

HOW MANY ENTRIES WITH (COUNTRY,Peru).AND.(LIFE-FORM,Herb)*

    2 OUT OF   6 ITEMS HAVE THESE ATTRIBUTES (33.33 PER CENT).

PRINT:(ACCESSION NUMBER, BINOMIAL, ALTITUDE) FOR ENTRIES WITH RESULT*

    2 OUT OF   6 ITEMS HAVE THESE ATTRIBUTES (33.33 PER CENT).

1 C.incana   200 m.
6 C.nitens   1500 m.

PRINT:SAME FOR ENTRIES WITH .NOT.COUNTRY,Peru*

    4 OUT OF   6 ITEMS HAVE THESE ATTRIBUTES (66.67 PER CENT).

2 C.incana   850 m.
3 C.agatiflora   2600 m.
4 C.nitens   2100 m.
5 C.pilosa   2000 m.

of physically resorting the cards is tedious and time-consuming. Computer programs can do the sorting job with varying degrees of ease. A system which has maximum flexibility of access is likely to be slow but allows the user to access and rearrange the items by way of any descriptor, or combination of descriptors, used as the selection criterion. Alternatively, a system may be restricted to retrieving or indexing only on a fixed set of descriptors—perhaps the species name or certain label data—a fast but more rigid system. Note that in either case any amount of data may be requested for the output list; it is the selection criteria that may be limited.

The fully flexible, multiple-access capacity in TAXIR permits subsets to be extracted according to any selection criteria, taken singly or in combination. This allows one to extend the objectives of retrieval from recapture of data to simple analysis. For example, the computer is able to prepare lists of those items that fit certain descriptions. These can be used to suggest correlations when the lists or tallies contain the states of substantive descriptors, rather than simply identifiers. Thus, if all of the members of a subset defined by (FLOWER COLOR,Red) also list (CHEMICAL A,Present), an association is suggested. Use of the query under these conditions allows an easy, quick means for preliminary comparison and analysis of items within the data bank. It can serve as a means for checking preliminary hypotheses or "hunches" and, indeed, used as a scientist's "scratch pad."

Information is retrieved by searching the database. In most database management systems the data are subjected to sequential, or linear, search. Conceptually, the method of sequential searching is just that. The computer compares data entries with a criterion character set and "flags" the matches which then are pointed to an output list. Because the computer does simple comparisons rapidly—and tirelessly—such repeated comparisons are feasible. Furthermore there are now clever ways of programming and file management that reduce the length of the search. Still sequential searching is a rather slow process when large databases, or more than one database, must be searched.

Information retrieval in TAXIR operates without sequential searching of the database. The tabular system, combined with the rule that the table contain mutually exclusive, nonoverlapping states, allows both storage and retrieval in TAXIR to operate on different principles, which make it very rapid and efficient. This is because computers are essentially Boolean logic machines and the TAXIR algorithm, likewise, is based on set theory and Boolean algebra (Estabrook 1966). The program derives its efficiency from an ingenious means for reduction of the data to binary arrays, cued to the order of the descriptor states, which allows for dense packing of the information in memory. Even more importantly, retrieval in TAXIR is extremely rapid because each query can be answered by a Boolean calculation on the binary arrays—which a computer does very quickly—rather than by a linear search.

Searching Multiple File Databases

Up to now we have dealt with searching single tabular data structures (flat files) either sequentially or algorithmically. These two-dimensional files are suitable for some of the tasks in the real world of taxonomic information management. In practice, however, most taxonomic databases can not be limited to single flat files. To do so would require an initial permanent designation of an item that would become the basic element for the entire database. In a Flora, for example, one is likely to have data on specimens and also on populations, and the two are not consistent with respect to item designation. The same is true of a monograph, where it is logical for the items to be taxa, and there is related bibliographic data. In the latter the items are references described by author, title, etc. The fact is that most taxonomic information systems of any size contain **multiple files,** in the form of several related flat files. For such databases to be fully useful, information retrieval must allow searching on more than one file by means of a single query. In general terms, the capacity for **merging files** is needed.

Computers are designed for manipulating data files. Dealing with multiple files requires that the relationships among these files be specified. The several different kinds of relationships among data files that exist are known as **database structures.** These structures fall into three general classes (Barron 1984): (1) the **hierarchical** structures, in which connections among files form trees having a single point of entry or root; (2) the **network** structures, which allow any of the possible pairwise relationships among files; and (3) the **relational** database structures, in which related files are linked by means of the data they hold in common. There are database management systems based on all three of these models, which we will describe with some examples.

Hierarchical Database Management. People commonly store data in their memories in hierarchical form; biologists are especially familiar with this approach. A systematist collects many specimens and categorizes them, possibly according to species. Pertaining to each of P species there may be a set of data about the collection such as the location for each of the specimens. Let us say there are n_1 of these descriptors. There is, possibly, another file of data on n_2 morphological measurements on certain specimens within each of the P species. The species themselves may be grouped within a smaller number of genera. Hierarchical databases use exactly the same logic to arrange data in computer memory. The data are stored in a sequence of categories, some subordinate to others, and the program allows systematic searches through the files. A query functionally similar to the one described above is used as the instruction command for the search.

There is an example in fig. 11.6 that shows the result of a query applied to an hierarchical data structure in the GIS program (Krauss 1973). The data consist of a series of species that forms the primary set of items for this database. Under each species there are two subordinate files containing descriptive, multiple-entry data about each species. That is to say, for example, many states or provinces are likely to be appropriate descriptor states for any given species. In addition there are some parallel descriptors, such

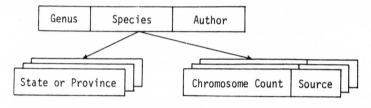

```
QUERY TAXON
IF STATE EQ 'MD' OR 'VA'
LIST TAXNAME, AUTHOR, CHROMCNT
EXHAUST TAXON
```

Taxon Name	Author	Chromosome Count
DIGITARIA VALIDA	STENT.	18
TRISETUM SPICATUM	RICHT.	28
CAREX ATRATA	L.	54
PALISOTA HIRSUTA	K. SCHUM.	40
ALLIUM GIGANTEUM	BAKER	16

Figure 11.6 Example of a query in a hierarchic database, GIS *(From Krauss 1973. Composite from Figs. 2, 4, and 5).*

as genus or author (of species name). These contain simple, single-entry descriptive data for each species.

All files can be subjected to linear or sequential searches. In the case shown in fig. 11.6, the query directs a search of the STATE OR PROVINCE file to pick out the subset that consists of items from either MARYLAND or VIRGINIA. For each of the items so selected the name of the SPECIES is held, the appropriate AUTHOR name, also from the primary file, is appended, and the CHROMOSOME COUNT information, stored in another subordinate file, is retrieved using the species name as a key. These three records for each of the selected items then form the output list that appears in catalogue form with appropriate headings.

We note that in the process of querying, two files—SPECIES-AUTHOR and CHROMOSOME COUNT—are merged. It is possible to store the result as a new file in the computer memory as well as to print an output list. Creation of such new files is a good move if it can save the time-consuming process of file search at some time in the future. The art of querying such a hierarchical database management system efficiently comes in managing files so that multiple useful cross reference files are maintained. This allows more rapid searches through mostly relevant data and limits the number of fully sequential file searches that must be made. When storing copies of merged files, caution must be taken to see that the new file is updated, or purged, when corrections are entered into the original data files. Otherwise the system will contain increasingly divergent versions of similar information, leading to different responses depending on how a question is formulated and presented.

The hierarchic database structure deals particularly efficiently with the problem of within-taxon variation. The variable descriptors are recorded only at the level in the hierarchy where they have ceased to vary! This saves space and also eliminates logically necessary but often biologically meaningless terms like 'not applicable' and 'variable.' For example, states of the CHROMOSOME COUNT descriptor, which is variable within genera, are stored only at the species level, where they can usually be indicated by a single number. By contrast, in a tabular format with genera as the items the information would have to be indicated by a range of numbers or by the expression 'variable' or by some textual listing of chromosome counts for the different species.

Network Database Management. Connecting links in a hierarchical structure are a subset of those possible in a network, or plex, structure in that the former are directed graphs (rooted trees). Networks allow "many to many," as well as the "one to many" linkages available in the tree structure. Network database management systems have been devised principally for mainframes and were until recently the most common system used commercially. Needed connectors among the data files must be designated and an instruction devised for "navigating the database." A data manipulation language related to COBOL is often used for this required programing and the term 'CODASYL database management systems', which is sometimes applied to network systems, derives from this relationship. The design of network systems tends to be implementation-oriented; that is, there is not total independence between the logic of the data retrieval scheme and the physical organization of the data in the computer (Martin 1975).

Relational Database Management. The programing and program language requirements for searching network databases make them somewhat difficult to use for those without extensive computer training. At present, the relational approach appears to be a more promising one for these users. A major advantage is that one does not have to pre-define the structure of the retrieval when using a relational database. Furthermore, relational database systems are now available for microcomputers, and this adds to their appeal for taxonomists.

The idea of a relational database is easy to understand because it is related to common sense. Manipulation of relational databases is based on the mathematical notions of set theory and matrix algebra, which has advantages that affect computability but do not need to concern users. Users can regard the relational database as a series of tabular data files that are linked implicitly because two, or more, tables contain the values of a descriptor common to both. This is to say the data tables are linked by their contents.

The data tables in a relational database are known as **relations.** Relations can be manipulated using the following three basic operations:

SELECT, in which an item (record) is extracted from the relation.
PROJECT, in which descriptors may be rearranged and deleted and duplicate
    items removed.
JOIN, in which tables are combined where descriptors have common values.

The use of a series of these operations, in accordance with the requirements of a query, results in a sequence of new relations such that the last one contains the data requested. Let us look at a very simple (fictitious) example. Three data tables—relations—are shown in fig. 11.7. They all deal with the collection and distribution of seeds grown for different species of *Vicia*. The first relation holds data on the plants that produce the seed, the second on the seeds collected, and the third on the seed shipments. Because relational database management systems always use columns for descriptors (termed domains) and rows for items (rows are elements in the set of *n*-tuples that

Plant relation

PLANT NO	SPECIES	ORIGIN
P1	Vicia faba	Unknown
P2	Vicia ervilia	France
P3	Vicia lutea	Spain
P4	Vicia ervilia	Portugal
P5	Vicia ervilia	Turkey
P6	Vicia sativa	UK

Seed Packet relation

PACKET NO	PLANT #	QUANTITY	YEAR
001	P1	50 g.	1971
002	P2	40 g.	1972
003	P2	100 g.	1972
004	P3	100 g.	1973
005	P1	50 g.	1973
006	P5	30 g.	1974
007	P6	70 g.	1974
008	P1	40 g.	1975
009	P3	30 g.	1977
010	P4	90 g.	1977
011	P5	50 g.	1980
012	P3	50 g.	1980

Shipments relation

PACKET #	RECIPIENT	DATA
001	ABB	1/9/72
002	BIS	2/7/73
003	ROG	3/1/73
004	ABB	1/2/75
005	ABB	1/9/75
006	BIS	8/17/76
007	ROG	3/14/75
008	ABB	5/11/75
009	ROG	9/12/79
010	BIS	8/7/80
011	ABB	9/2/81
012	ABB	1/14/82

Figure 11.7 Sample relational database tables.

comprise relations), we shall abandon, for the moment, the opposite convention, which we have used throughout this book.

Data from all three relations would be required to answer the
query:

Print: SPECIES ORIGIN, for those shipped to ABB

From the Shipments relation, the relational database management program
would SELECT and form a new relation containing the following:

PACKET #	RECIPIENT	DATE
001	ABB	1/9/72
004	ABB	1/2/75
005	ABB	1/9/75
008	ABB	5/11/75
011	ABB	9/2/81
012	ABB	1/4/82

1 (the number 1)

A JOIN with the Seed Packet relation, on the basis of the 'Packet Number'
descriptor, leads to the following relation:

YEAR	QUANTITY	PLANT #	PACKET NO	RECIPIENT	DATE
1971	50 g.	P1	001	ABB	1/9/72
1973	100 g.	P3	004	ABB	1/2/75
1973	50 g.	P1	005	ABB	1/9/75
1975	40 g.	P1	008	ABB	5/11/75
1980	50 g.	P5	011	ABB	9/2/81
1980	50 g.	P3	012	ABB	1/14/82

1 (the number 1)

Using PROJECT to delete columns QUANTITY, DATE, RECIPIENT,
PACKET NO, and YEAR, and JOIN to attach data from the Plant relation
leads to the relation:

PLANT NO	SPECIES	ORIGIN
P1	Vicia faba	Unknown
P3	Vicia lutea	Spain
P5	Vicia ervilia	Turkey

The result now contains the information that was requested in the query, and
it could be printed out. Note that the repeated instances of numbers 1 and 3
plants, which would have resulted in duplicated rows in the last relation,
have been removed, a function of the PROJECT operation.

Choice of Database Management Systems

It is obvious now that utilizing a database management system will require making many choices at the start of the project. The first decision should be based on the type and structure of data that the project will involve and the kind of analysis that is required. Can the data be arranged in a single flat file? Or will there be several related files? If so, are they hierarchical or are there other relationships among them? Will using the data require extensive analysis, such as character and cluster analysis in a classification project? Or is the major purpose to store data in retrievable form, such as in a curatorial or descriptive monographic project?

The answers to these questions concerning the data in and the purposes of the project should play a large role in the choice of a database management system. For instance, if the data form a single flat file and if a cluster analysis is to be done, then a simple file management program with an easy-to-use, rapid querying capacity might be a good choice. If the data structure is essentially hierarchic, then a flat file management system is very awkward to use and one must choose instead a database management system that allows easy and rapid merging of files.

Factors like these then influence the choice of software and hardware. For the user with an application in mind the basic decision is the choice of a program—software—that will suit the project's data and analysis needs. However, especially for micros, software and hardware are not entirely separable. In many cases the choice of a micro might be made in light of the program that is to be run.

Besides appropriateness for the data and the analysis and general "friendliness"—that is, an easy to use program that will indeed be used by workers on the project—there are a number of more external considerations that need to be kept in mind. It is usually best to choose a system that is reasonably widely used so that software and operating systems that have been well tried are readily available. The opposite extreme of having a hand-built system, while it has the potential advantage of being exactly tailored to the user's needs, usually causes problems. The system is apt to become dependent on an in-house programmer and there may be a disaster when that person departs.

For all but the most limited use database management systems need to be programmable. This allows adaptation and customization to the needs of several workers in a project group. For example, specialized programs may be constructed for frequently repeated routines.

Expansion of the goals and activities of projects should be allowed for in planning the database management system. This is especially relevant to early decisions about systems that will handle flat files versus hierarchical data. Even a curatorial database, which may at first appear to be an ideal candidate for a simple file management program, can expand in such a way that simply enlarging the file or appending a second file is not enough. There may be a need to interrelate the databases—a task for a relational or other multiple file system.

In the last analysis, however, no matter how much time and energy have been put into planning a computerized database management system, the probability is that it will one day become outmoded. At the present time the burgeoning computer software industry is marketing software with major improvements as often as every six months. This is not to say, of course, that the user must always have the very latest program. On the contrary, once a system has been adopted and used for a project the investment of time, as well as money, in it is sufficient that it is reasonable to stick with it for a period of perhaps five years. By that time the micros are likely to have worn out anyway, and the time will be ripe for a fresh consideration of the new developments in database management systems.

Suggested Readings

General

Allkin, R. and F.A. Bisby 1984. *Databases in Systematics*. London and New York: Systematics Association, Academic Press.

Computers

Date, C.J. 1981. *An Introduction to Database Systems*. Menlo Park, California: Addison-Wesley.
Martin, J. 1975. *Computer Data Base Organization*. Englewood Cliffs: Prentice Hall.

Chapter 12

Taxonomic Databases

I t is quite apparent by now that research in taxonomy and systematics both generates and requires large quantities of data. Up to now we have dealt mainly with descriptive data, that is, the descriptors and characters that define specimens and taxa. We have shown how such data can be organized into two-dimensional tables, or into databases; how data can be analyzed to produce taxonomic products, classifications, keys, and retrieved by querying. Taxonomists use many other types of databases, and these are often combined into a system, which we shall refer to as an **information system.** Taxonomic information systems are composed of a number of separate but related databases. For instance, the biogeographic data for the distributions of the native species of the genus *Vicia,* the bibliographic data on the family Leguminosae, and the list of the valid names and synonyms of the Leguminosiae tribe Vicieae are all examples of databases. They are related in that they all deal with some aspects of the legumes. Thus, they are all possible elements of the same information system.

Types of Databases

Databases that belong to an information system can contain very different types of information, as these examples illustrate. The structures of these databases, members of the same information system, are also different. For example, the items forming the basis for the geographic database would be species while that for the bibliographic would be publications or authors. As we have seen, there are database management methods that enable us to store the data in several databases efficiently and still use it in an integrated fashion.

We classify databases into several types depending on their function and the biological meaning of their contents. The major types are the following (Bisby, 1984b):

> *Curatorial.* Accession data for items in collections as found in museums, herbaria, botanic gardens, germplasm collections, and zoos.
> *Biogeographic.* Distribution data for taxa as found in atlases, and in inventory, distribution, and conservation projects.
> *Bibliographic.* Bibliographic details of publications relating to the taxa.
> *Nomenclatural.* Names, sometimes with details of authority, publication, and typification, and often containing valid names cross-indexed to synonyms.
> *Descriptive.* Descriptions of specimens or taxa such as details of morphology, anatomy, cytology or chemistry.

The **curatorial databases** found in large institutions, where they play a major part in the management of the collection, are typically large but simple in structure. They contain many items and relatively few descriptors. Several examples will be given later in this chapter. In addition there are numerous small curatorial databases that are designed to assist an individual or a group working on a monographic project or Flora or Fauna. They may, for example, provide means for organizing specialized collections or for indexing sources and locations of data and names.

Biogeographic databases can be a result of the analysis of data arising from the collections that are made in a large floristic or monographic study. In one case (fig. 12.1) the result is a list of species with indications of the geographic areas where they have been found. In another (fig. 12.2) the data may be published in the form of maps indicating the locations where a particular species has been found within a given geographical area.

Bibliographic databases, like curatorial databases, may be very large, even worldwide, in scope; or they may be small and related to personal goals or to those of a research group. In either case bibliographic databases contain references to the literature pertaining to the taxa under study. On the largest scale, BIOSIS PREVIEWS, published by BioScience Information Service (BIOSIS Inc.) of Philadelphia, meets the needs of taxonomists as well as those of other biological scientists. BIOSIS PREVIEWS is available in 62 countries as an online computer database. In addition there are subscription search services such as B-I-T-S (Schultz, 1983) which produces selected sets of references, with or without the published abstracts, tailored according to the research interest of the subscriber. Subscribers regularly receive the information, on floppy discs that are compatible with their

Table 1.   **Continental Distribution of *Vicia***   Table 1.

Name	EUR	ASI	AFR	N.A	S.A
Vicia ciceroidea	-	P	-	-	-
Vicia ciliatula	-	P	-	-	-
Vicia cirrhosa	-	-	P	-	-
Vicia cracca	P	P	P	-	-
Vicia cracca subsp. atroviolacea	P	P	-	-	-
Vicia cracca subsp. cracca	P	P	-	-	-
Vicia cracca subsp. gerardii	P	-	-	-	-
Vicia cracca subsp. stenophylla	P	P	-	-	-
Vicia cracca subsp. tenuifolia	P	P	P	-	-
Vicia cretica	P	P	-	-	-
Vicia cretica subsp. aegaea	P	-	-	-	-
Vicia cretica subsp. cretica	P	P	-	-	-
Vicia crocea	-	P	-	-	-
Vicia cuspidata	P	P	-	-	-
Vicia cypria	-	P	-	-	-
Vicia dadianorum	-	P	-	-	-
Vicia dennesiana	P	-	-	-	-
Vicia dichroantha	-	P	-	-	-
Vicia disperma	P	-	-	-	-
Vicia dumetorum	P	-	-	-	-
Vicia durandii	P	-	P	-	-
Vicia epetiolaris	-	-	-	-	P
Vicia ervilia	P	P	P	-	-
Vicia esdraelonensis	-	P	-	-	-
Vicia faba (Wild origin unknown)	-	-	-	-	-
Vicia faba subsp. faba (ditto)	-	-	-	-	-
Vicia faba subsp. paucijuga (ditto)	-	-	-	-	-
Vicia filicaulis	P	-	P	-	-
Vicia floridiana	-	-	-	P	-
Vicia freyniana	-	P	-	-	-
Vicia galeata	-	P	-	-	-
Vicia galilaea	-	P	-	-	-
Vicia galilaea subsp. faboidea	-	P	-	-	-
Vicia galilaea subsp. galilaea	-	P	-	-	-
Vicia glareosa	-	P	-	-	-
Vicia glauca	P	-	P	-	-
Vicia graminea	-	-	-	-	P
Vicia grandiflora	P	P	-	-	-
Vicia hassei	-	-	-	P	-
Vicia hirsuta	P	P	P	P	-
Vicia hirticalycina	-	P	-	-	-
Vicia hugeri	-	-	-	P	-
Vicia hulensis	-	P	-	-	-
Vicia humilis	-	-	-	P	P

KEY TO ABBREVIATIONS

AFR	Africa	N.A	North America		P	present
ASI	Asia	S.A	South and Central America		-	absent
EUR	Europe					

Figure 12.1 A distribution list from a biogeographic database. *(From Allkin et al 1983b. p. 6 of Table 1).*

Especímenes Examinados:

Balls 4701 (A)
Botteri 262 (F)
Dorantes & Acosta 2045 (ENCB)
Gómez-Pompa 4353 (GH, MEXU)
Hernández-Cerda 4 (MEXU)
Hernández y Trigos 808 (F, MEXU)
Matuda 1230 (A, MEXU, NY)
Miller 1340 (GH)
Nevling & Gómez-Pompa 2171 (F)
Pringle 8199 (F, MEXU)
Sharp 44661, 45615, 46158 (MEXU)
Sosa 47 (INIREB, MEXU)
Vela y Martínez 1275 (ENCB)
Ventura 1328 (ENCB, F); 8283 (ENCB); 10589 (ENCB, MEXU)
Weaver, Foster & Kennedy 1714 (F)
Zolá 703 (F, INIREB)

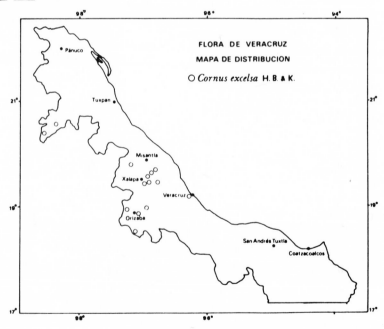

Figure 12.2 Computer generated species distribution map. *(From Sosa 1978. Fig. 1, p. 6).*

particular microcomputer and operating system. There is also special software available to allow the subscriber to create, update, and search the file of references that is being accumulated. Besides those from *Biological Abstracts* there are other large online bibliographic databases of which CAB Abstracts from the Commonwealth Agricultural Bureau (CAB) and AGRICOLA from the United States Department of Agriculture (USDA) are the most important.

In monographic work the references in the literature to the taxa under study are an important data source. An example of the usefulness of computerizing bibliographic data is seen in some work one of us (DJR) did on the genus *Vitis*. For each species bibliographic data and items of nomenclatural importance were recorded. The computer program (TAXIR) enabled him to produce lists containing authorities, dates, and synonyms like that shown in fig. 12.3.

Print: [Publ Name; Auth; Species] for with item; from 1 to 209 *

No. of items in query response = 209.
No. of items in the data bank = 209.
Percentage of response/total data bank = 100.00.

Addr. on Amer. Grapes	Simpson, J. H.	V. munsoniana
Allg. Gartenzeitung	Otto, R. + Dietrich, A.	V. isabella
Amer. Midl. Naturalist	House, H. D.	V. lecontiana
Amer. Midl. Naturalist	House, H. D.	V. shuttleworthii
Amer. Naturalist	Engelmann, George	V. arizonica
Ann. Mus. Bot. Lugduno-Batavum	Miquel, F. A. W.	V. pubescens
Ann. Nat.	Rafinesque, C. S.	V. denticulata
Arbust. Amer.	Marshall, Humphry	V. laciniosa
Arbust. Amer.	Marshall, Humphry	V. vinifera
Beitrage Naturk.	Ehrhart, Friedrich	V. hederacea
Berl. Baumzucht.	Willdenow, Carl L.	V. indivisa
Bio. Centr.-Amer. Bot.	Chamisso, L. A. et Schlectendal, D. F. apud Hemsley	V. elliptica
Bio. Centr.-Amer. Bot.	Decandolle, A. P. apud Hemsley	V. tuberosa
Bio. Centr.-Amer. Bot.	D. C. apud Hemsley	V. mexicana
Bio. Centr.-Amer. Bot.	Humboldt, Bonpland + Kunth apud Hemsley	V. tiliacea
Bol. Esp. Hist. Nat.	Fragoso, R. G. + R. Ciferri	V. bellidifolia
Boston J. Nat. Hist.	Engelmann, George In A. Gray	V. candicans
Boston J. Nat. Hist.	Lindheimer, F. J. In A. Gray	V. populifolia
Bot. Beecheys Voy.	Hooker, W. J. + G. A. W. Arnott, Decandolle apud	V. caribaea quest.
Bot. Cultivateur, Ed. 2	Dumont de Courset, G. L. M.	V. cordata

Figure 12.3 Lists from a bibliographic database. *(From Rogers and Rogers 1978. Fig. 3).*

As the last example suggests, **nomenclatural databases** are related to bibliographic in that an initial literature survey often includes a search for accepted names, synonyms, locations of type specimens, authorities, and other nomenclatural details. At its conclusion a monographic project may produce one or more nomenclatural databases. In the Vicieae database project, which is going on in the laboratory of one of us (FAB), the names and synonyms of species and subspecies of the tribe have been published in the form of a pamphlet (Allkin et al. 1983), which contains simply a list of the 302 accepted names for members of the Vicieae tribe. Each is arbitrarily numbered and followed by its list of synonyms. The list is generated by a computer program (SYNONYMS), which uses a file that is separate from, but substantively related to, the general database for the project (see fig. 12.13 below).

Descriptive databases vary with respect to structure as well as substance. If they contain information that is pairwise, such as that for hybridization trials among subspecies, the structure is different from that, say, for pedigree data. Some descriptive information, such as chemical data, may have a fairly simple tabular structure. For example, a chemotaxonomic database for the Vicieae tribe of Leguminosae contains information on 85 substances in 111 species (Babac and Bisby 1984). Although there are multiple records per species, to accommodate information on plant parts such as seeds and leaves as well as the whole plant, the database is a simple flat file having two or three descriptors per substance. It contains information on presence-absence, and in some cases quantities of the substances, as well as literature citations for the sources of the information.

The morphological descriptive information, which is central to taxonomic work, is also the most complex to organize into useful databases. Some of this complexity is biologically based. It is important to note, first, that there is a distinction between features or characteristics of the organism and descriptors or characters in the data structures (Allkin 1984) and, second, that some biological features result from a combination of a number of genetically based traits.

These features do not, therefore, provide characters that are single bases of comparison but instead generate a number of characters. As an example, take the case of inflorescences (Pankhurst 1984). Inflorescences are the flower clusters present on flowering plants. The structures of these arrangements of flowers are highly variable and complex, and traditionally have been described by technically defined, overall terms like raceme, panicle, spike, umbel, capitulum.

Unfortunately, these terms subsume such a wide range of variability that there are cases where none of them fit precisely, as is true for

Geranium molle. In order to make characters better suited to computerized keymaking, Pankhurst has devised a list of 12 component characters, such as first- or second-order branching, presence of flower stalks, or terminal flowers on main axes. These apply to all of the inflorescence types. Thus, by dividing a single complex feature into a number of characters we are assured of a set of comparable characters with nonoverlapping states.

The character by taxon tables that we have used in our illustrations of taxonomic analysis up to now represent the data organization that is ideal for purposes of data manipulation and retrieval. As we have said the tabular structure promotes the use of standardized terminology; it demands comparable and complete data; and when single entries per cell are the rule, it assures nonoverlapping, mutually exclusive character states. However, it is not always possible to force the descriptive taxonomic data into such a simple tabular format. Difficulties occur that have to do both with the substance and the structure of information about living organisms.

The structure of the data concerning taxa is inherently hierarchic. For instance, a database with species as the items will always hold information that pertains to variations among the species in explicit form. Descriptive information that is invariant among the species within a genus is usually contained in the database only implicitly, via genus descriptions. It can be made explicit and entered into the tabular data structure by adding variables and repeatedly entering the variable states held in common by the species of each genus. However, this technique rapidly becomes cumbersome, particularly when many levels of taxa are involved such as in the database for a Flora or Fauna.

Another source of difficulty in organizing descriptive data is due to dependencies among characters. Frequently, groups of specimens or taxa do not have completely comparable characters because one or more of the characters depends on the presence of some other character; in its absence one or more following variables can be recorded only as 'inapplicable'. See fig. 12.4 for an example. Here again, the difficulties increase greatly as the items become more variable and the problem is most acute with large databases planned to accommodate, for example, the flora of a region.

There are some places in taxonomic work where the items of the database are specimens—perhaps in the initial stages of a monographic research project or in a curatorial database. Very frequently, however, the taxomist works with taxa and the items in a tabular database are, thereby, populations. There is inevitably internal variability with these groups, and problems arise in recording such data. One must either record more than one entry per cell or more than one description per taxon. An ideal structure would enable us to include estimates of the frequencies of variable descriptor states as well

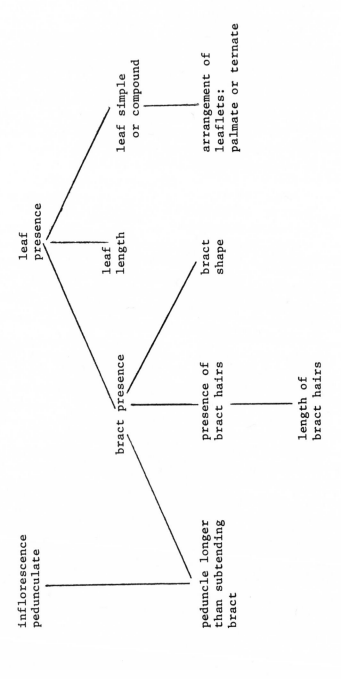

Figure 12.4 Illustration of character dependencies. *(From Allkin 1984. Fig. 1).*

as to identify joint character state distributions among taxa. Unfortunately, such ideal data-handling systems have yet to be devised.

A different kind of problem in handling descriptive data arises from the propensities of taxonomists, like other scientists, to record data in field notes or in laboratory notebooks in textual format. Such notes contain maximal amounts of information, but in a form that is difficult to handle for a number of reasons. It frequently contains qualifiers and comments that can be interpreted only by the investigator. It is likely to be incomplete and non-comparable, and to contain much implied information. However, comments and unformatted text elements are often necessary parts of descriptive data, inconvenient as they may be to organize.

An information-storage system for taxonomic descriptions has been developed by the group at the Australian National University and CSIRO, Canberra (Dallwitz 1980; Macfarlane and Watson 1982). It is especially designed to code the information in descriptions of the textual type found in monographs. The coding program goes by the name DELTA, acronym for DEscriptive Language for TAxonomy. The purposes are several. In the first place, the presence of a preset system for coding facilitates data recording and helps to ensure consistent and complete data. Secondly, the coded descriptions are better suited to computer storage in that they are easily manipulated for conversion to other formats and for collecting or reordering subsets of the items.

Coding is accomplished by numbering both characters and character states. Thus, the first step is to designate and number the character names and the character state names that will be used. In addition to this essential data, comments may be used to include incidental modifiers and subsidiary material. This type of textual data is separated and identified by its enclosure in angle brackets (⟨ ⟩). Each character name or state is ended by a solidus (/) followed by a blank and the data are entered in free format using these separators, as seen in the example in fig. 12.5 from Dallwitz (1980). Any of the usual character types—like ordered or simple multistate, integer, or real numeric—are recognized and do not require particular designations or groupings.

An item description for each of the taxa begins with the taxon name, with any desired comments in brackets. Next the taxon is described using the character number as its coded name followed by a comma and then by the number of the state that applies. Unlike the case in the strictly tabular, nonoverlapping format, there may be more than one appropriate character state per character. Such multiple states are separated by the solidus following each number which is read as *or*. Character states may also be connected by "—" meaning *to* or "&" meaning *and*. Again any desired comments may be added in brackets. The coded data for the items is written

#1. Striated area on maxillary palp/
    1. present/
    2. absent/
#2. Pronotum <colour>/
    1. red/
    2. black/
    3. yellow/
#3. Eyes <size>/
    1. normal/
    2. very large/
#4. Frons/
    1. with setae on anterior middle and above eyes/
    2. with setae above eyes only/
    3. without setae/
#5. Number of lamellae in antennal club/
#6. Length/ mm/

Figure 12.5 Examples of coded character descriptions. *(From Dallwitz 1980. Table 1, p. 42).*

in free format, the different characters being separated by blanks. Thus, the following code from Dallwitz (1980):

"1,1/2 ⟨rare⟩ 2,2/2&3 ⟨striped⟩ 3,1—2 6,7—8.5"

stands for:

"striated area on maxillary palp; or absent (rare).
Pronotum black; or black and yellow (striped).
Eyes normal to very large. Length 7 to 8.5"

A major advantage of DELTA coding of taxonomic descriptions is that the data can be readily reformatted. Thus, it can be made available in the forms required for other programs. Examples are KEY, which produces keys automatically (Dallwitz 1984) and MULTBET, which does classifications. Conversion of stored data into the required format can be done by a program known as CONFOR.

CONFOR can also be used to translate the coded data back into natural language as fig. 12.6 illustrates. The descriptions generated by CONFOR sometimes appear to be clumsily verbose due to the repetition of character names and character state names for each element of the description. However, the pleasure of finding truly comparable descriptions easily compensates for the tedium of reading the formal output. Note that the character numbers are also inserted; they can greatly facilitate comparisons of character states among taxa. However, it should be noted that the CONFOR program does not provide retrieval capacity in the sense of a querying capability for extracting subsets of items having designated character states.

Automated typesetting is available for the CONFOR output. Since both upper and lower case can be used and a wide variety of type fonts are available, retyping and repeated proofreading is largely eliminated. The program TYPSET either inserts print instructions or interprets those put into the original data by the user. The letter 'B' next to 'Neyraudia' in the example (fig. 12.6) is the instruction for bold face printing of the name.

Since the typesetting device used is a phototypesetter, the images can also be reproduced as microfiche. These are cheap enough to permit replacement after revisions are made in the database and transmission through the post.

A large application of the DELTA—CONFOR—TYPSET system is a book on Australian genera of grasses (Watson and Dallwitz 1980). It contains nearly 200 pages of descriptions and keys produced automatically from the database. Similar books could be produced for other regions from the same database. Revisions to the monograph are produced on microfiche at regular intervals.

Examples of Databases

Since databases of different types function in ways that serve the various aspects of taxonomic inquiry, the databases actually used in taxonomic projects, expecially those that are large-scale ones, will be of more than one type. The set of examples below will illustrate this; other examples are to be found in Allkin and Bisby, 1984 and Sarasan and Neuner, 1983. We shall first discuss the use of various computer programs in a number of curatorial projects: the herbaria at the University of Colorado and the University of Michigan, the museum at the University of Kansas, the Colombia National Herbarium, and a museum documentation system that serves many museums in the United Kingdom. Then, a number of major floristic and faunal projects employing computerized database management will be discussed: Flora Veracruz, which deals largely with a specimen collection, and the European Documentation Service (ESFEDS) which is now developing biogeographic databases and plans major nomenclatural and bibliographic services relating to European vascular plants. The latter will be considered along with two other bibliographic projects, Index Kewensis and the international BIOSIS Taxonomic Reference File (TRF). Lastly we shall describe two representatives of large-scale monographic and phytogeographic projects. These in-

\# [BNeyraudia] <Hook. f. 2 tropical Africa, Madagascar, China, Indomalaya>/
2,2 3,1<cf. [IArundo]> 5,2 6,V 7,75–500 9<forming loose tufts> 10,1 13,1 16,2 17,1 18,2 20,2 21,2 24,2 25,1<1.5[X–]4 cm, often longitudinally splitting> 26,2 27,2 28,2 29,2 30,1 32,1 34,3<long> 36,– 37,2 38,2 40,2 41,2 42,1 43,2 45,1 46,5<large> 47,1 48,– 49,– 50,2<more or less verticillate> 51,3 52,2 53,2 54,– 55,– 56,– 57,1 58,2 59,– 60.– 61.– 62,– 63.– 64.– 65,2 67,2 68,– 69,1 70,– 71,2 72.2<thin pedicels> 73,2 74,2 75.– 76,– 77,1 78,5–9 79,– 81,1 82,1 83,1 84,1<but glabrous except above the abscission zone of each floret. The conspicuous hair tuft is on the "callus"> 85,2<somewhat below the lemmas> 86,1<long white hairs> 88,1 89,– 90,2 91,2 92,2 93,2 94,1<acute to acuminate> 95,1 96,2 97,2 98,1 99,2 100,V 101,– 102,2 103,2 104,3 105,1–3 106,1–3 107,4–8 108.– 109,2 110,1 111,3 112,1 113,1 114,2 115,2<acuminate, larger than but similar to the glumes> 116,3 117,1 118,2 119,2 120,2<membranous> 122,2 123,3 124,1<the distal florets with longer awns> 125,1 126,1–2 127,1/3<the lateral nerves may be excurrent into very short awns> 129,– 130,1 131,– 132,1<curved> 133,1<bearded marginally and on lateral nerves> 134,2 135,– 136,1<hyaline> 137,2 138,3 139,1 140,2 141,1 142,2 143,1 144,1 145,2 146,1 147,2 148,2 149,1 150,2 151.– 152,2 153,3 154,1 155,2 156,2 157,1 158,2 159,2 160,2 162,1 163,2 164,2 165,1<1.5[X–]2 mm> 167<grain angular> 170,– 188,2 189,– 190,1 191,1 192,2<the costals narrower and relatively longer> 193,1 194,1 195<and pitted> 196,2 197.– 198,– 199,2 200,2 201,2 202,2 203,2 204,2 205,2 206,2 207,1 208,2 209,2 210,2 211,1/2 212,1<mostly high domes> 213,2 214,1 215,1<and a few solitary> 216,1<crescentic silica bodies> 217,1<but many of the short-cells quite long> 218,1 220,2 221,2 222,1 224,2 225,2 226,1 227,1 228,2<bundle somewhat larger> 229,1 230,1 231,1 232,1 233,2 234,2 235,2 236,2 237,2 238,1<all bundles> 239,1 240,2<small groups of fibres occur abaxially under the bulliforms> 242<2n = 40> 243,12

Neyraudia <Hook. f. 2 tropical Africa, Madagascar. China. Indomalaya>.

(2) Perennial. (3) Culms woody and persistent <cf. *Arundo*>. (5) Culms self supporting. (6) <Height of mature plants, whether 3m or more> variable. (7) Mature culms reaching 75 to 500 cm. (9) <Forming loose tufts>. (10) Rhizomes pachymorph. (13) Culms branching vegetatively above. (16) Nodes glabrous. (17) Culm internodes solid, or spongy. (18) Fresh shoots not aromatic when crushed. (20) Plants unarmed. (21) Leaves not distinctly basally aggregated. (24) Leaf blades neither cordate nor sagittate. (25) Leaf blades broad <1.5–4 cm, often longitudinally splitting>. (26) Leaf blades not setaceous. (27) Leaves not needle-like, plants not prickly. (28) Leaves not pseudopetiolate. (29) Transverse veins very inconspicuous or invisible in the lamina. (30) Leaf laminae (or at least many of them) ultimately disarticulating entire from the sheaths. (32) Ligules consistently present. (34) Ligules consisting of a fringe of hairs <long>. (37) Leaf auricles absent. (38) Sheath margins free.

(40) Without pseudospikelets. (41) Plants bisexual with bisexual spikelets. (42) Having hermaphrodite florets. (43) Spikelets all alike in sexuality on the same plant.

(45) Inflorescence determinate (semelauctant). (46) Inflorescence paniculate <large>. (47) Overall form of panicle open. (50) Inflorescence neither digitate nor subdigitate <more or less verticillate>. (51) Spikelets not aggregated in heads, glomerules or spike-like units. (52) Inflorescence not leafy, not spatheate. (53) Inflorescence not comprising "partial inflorescences" and foliar organs. (57) Panicle branchlets capillary. (58) Not showing disarticulation of the spikelet bearing rachides. (65) Spikelets not all embedded.

Figure 12.6 (left) DELTA coded form of a genus description (above). Natural language description produced by CONFOR. Note that only 57 of the 240 characters in (a) appear in (b) demonstrating the condensation afforded by DELTA. *(From Dallwitz 1984. Figs. 2 and 3 respectively).*

clude the Vicieae Database Project at the University of Southampton (UK) and the Brassicaceae database at Notre Dame (USA).

A major monographic work that employs a highly automated printing system is the one on the grasses underway in Canberra, Australia (Watson and Dallwitz 1980). Since we have already described some of their work above in the section on descriptive data management, we shall not devote a separate section to it. However, it should be recognized as a major monographic database that illustrates well how a general database can be used for multiple purposes.

Before describing ongoing projects we should take note of the monographic project on *Manihot* (Rogers and Appan 1973, Rogers and Fleming 1973) which was the first demonstration of the applicability of computerized information management systems in taxonomic work. The project contained a coordinated series of three computer programs designed to assist taxonomists with their work. The first job—formulating the data table—was aided by the TAXIR information retrieval system. The next—that of determining useful characters and establishing proper character states—involved the use of the character analysis program CHARANAL as an aid. Lastly, the project was a major application of the early automated clustering program GRAPH. On the basis of the interpretation of the output of GRAPH, with a heavy emphasis on the using linkage diagrams to elucidate interrelationships among specimens, the complex of cultivars in *Manihot esculenta* (Rogers and Fleming 1973) was subdivided into related groups which have been found to be useful by agricultural workers and of interest to anthropologists.

The Universities of Colorado and Michigan Herbaria and the use of TAXIR

The TAXIR database management system was first invented at the University of Colorado (Estabrook and Brill 1969). It has been further developed at the University of Michigan (Brill 1983). Additions were made to a version named EXIR, at the University of Southampton, and there is a French version used at the University of Quebec (Legendre 1978).

As we have shown, TAXIR (EXIR) is especially well suited to flat files. Museum and herbaria collections are specimen based and thus amenable to tabular treatment. Fig. 12.7 gives an example of the use of TAXIR at the University of Colorado Herbarium in Boulder. Here it has been used to order information about specimens in the collection according to genus and species. We show just a small sample of the printout which forms a catalogue of more than 4000 vascular plants. TAXIR is also being used to

```
PRINT:  (GENUS,FAMILY),(SPECIES,PROVENANCE,HERB REC,NUMBER),(STATUS,LIT REF,RARITY),
(COLO REF,COMMENT) FOR ITEMS WITH NOT NUMBER,8000*

NO. OF ITEMS IN QUERY RESPONSE = 4398
NO. OF ITEMS IN THE DATA BANK = 4407
PERCENTAGE OF RESPONSE/TOTAL DATA BANK = 99.80

ABIES               PINACEAE
     CONCOLOR /GORD. ET GLEND./ LINDL.                                 IND          SPEC            1
          OK                                        HDH 1954                           NA
               HDH 1954                 ---
     LASIOCARPA /HOOK./ NUTT.                                          IND          SPEC            2
          OK                                        HDH 1954                           NA
               HDH 1954                 ---
     LASIOCARPA /HOOK./ NUTT. VAR. ARIZONICA /MERR./ LEMMON            IND          SPEC            3
          OK                                        HDH 1954                           NA
               HDH 1954                 ---
ABRONIA             NYCTAGINACEAE
     CARLETONII COULT. ET FISCH.                                       IND          SPEC            4
          OK                                        HDH 1954                           NA
               HDH 1954                 ---
     ELLIPTICA A. NELS.                                                IND          SPEC         4159
          OK                                        STANDLEY 1909                      NA
               HDH 1954                 ---
     FRAGRANS NUTT. EX HOOK. VAR. FRAGRANS                             IND          SPEC            5
          OK                                        HDH 1954                           NA
               HDH 1954                 ---
     FRAGRANS NUTT. VAR. ELLIPTICA HEIMERL                             ---          ---          4063
          SEE ABRONIA ELLIPTICA                     ---                                NA
               ---                      ---
     FRAGRANS NUTT. VAR. GLAUCESCENS A. NELS.                          IND          SPEC            6
          OK                                        HDH 1954                           NA
               HDH 1954                 ---
```

Figure 12.7 Part of a catalogue produced by TAXIR for a herbarium collection. *(Kindly provided by University of Colorado Museum, Boulder, Colorado).*

maintain a nomenclatural databank on flowering plant specimens at the University of Michigan Herbarium (Estabrook 1979).

The Museum of Natural History, University of Kansas and the use of SELGEM

An important database management system for museums that is widely used in the United States is SELGEM, an acronym for SELf-GEnerating Master. The program was designed and is maintained by the Smithsonian Institution in Washington, D.C. It is used to manage more than 100 files at that institution and by many others at more than 60 other museums. The policy is to make the system available free to nonprofit institutions in the U.S. and abroad.

The SELGEM system requires that a serial number be assigned to each item, and a three-digit code number to each category (equivalent to the descriptor in our terminology). The example in Fig. 12.8 shows how data on a single chipmunk specimen are stored at the Museum of Natural History at the University of Kansas in Lawrence, using SELGEM. Note the "preliminary listing" shown for the specimen: each line of the record, tagged with the serial number 131977, is an entry for a single descriptor which is identified by a category code like 051, 052, and so on. Thus, all entries per item are in the *(De, ds)* form. In our example there are 17 descriptors recorded for the specimen, including genus, species, subspecies, date, place of collection, collector's name and number, museum and accession numbers, sex, a set of physical measurements (category number 401, which includes five measurements: total length, tail, hind foot, and ear lengths and weight), and type of preparation.

Once entered and verified the descriptors in the preliminary listing can be reformatted and printed out in various orderings. The "final catalogue" entry for the same chipmunk specimen is also shown in fig. 12.8a. Note that all the descriptors are the same except genus name, which has been changed from "Eutamias" to "Tamias" because of a change in the generic status of the chipmunks. Such block corrections for a data field (genus, in this case) can be made easily in SELGEM. The SELGEM implementation in the Kansas museum is also designed to print out labels as the data are put in. Two examples are shown in fig. 12.8b—a skin tag and a skull box label.

Although SELGEM is probably strongest as a storage and museum data processing system, it also has a querying capacity and can be used for information retrieval in a more complex way than the simple reformatting needed for the catalogue lists and labels described above. Retrieval in SELGEM is based on sequential searches through single categories (descriptors). Sample query statements are seen in fig. 12.9. (The data are not the same as in the previous figures.)

```
00131977   051   01   KU
00131977   052   01   M
00131977   071   01   EUTAMIAS
00131977   075   01   MINIMUS
00131977   078   01   CONSOBRINUS
00131977   095   01   15 JUN 1972
00131977   100   01   U. S. A.
00131977   102   01   COLORADO
00131977   104   01   GARFIELD CO.
00131977   106   01   GLENWOOD SPRINGS, 17 MI N, 3 MI E OF
00131977   125   01   PATTERSON R R
00131977   126   01   2134
00131977   156   01   3463
00131977   200   01   1745
00131977   400   01   M
00131977   401   01   0172 0076  0029  0014  29.4
00131977   406   01   SKIN AND SKELETON
```

Preliminary Listing

```
131977    M    TAMIAS    MINIMUS    CONSOBRINUS
                                    U S A : COLORADO : GARFIELD CO
                                    GLENWOOD SPRINGS, 17 MI N, 3 MI E OF
                                    172 0076 0029 0014 29.4      SS     15 JUN 1972

                                                                3463
```

PATTERSON R R # 2134

Final Catalogue Form

UNIVERSITY OF KANSAS
EUTAMIAS   131977   M
   MINIMUS CONSOBRINUS
GLENWOOD SPRINGS, 17 MI N, 3 MI E OF
U. S. A.   COLORADO, GARFIELD CO.
15 JUN 1972   R. R. PATTERSON   2134
                1745
   MUSEUM OF NATURAL HISTORY

UNIVERSITY OF KANSAS MUSEUM OF NATURAL HISTORY
                       131977
EUTAMIAS MINIMUS CONSOBRINUS
1745   SKIN AND SKELETON

   131977   M   R. R . PATTERSON   2134
   GLENWOOD SPRINGS, 17 MI N, 3 MI E OF
   COLORADO GARFIELD CO.   U. S. A.
   172 76 29 14 29,4G   15 JUN 1972

Figure 12.8. (upper) Specimen data input for museum collection; (lower). Computer produced labels for the specimen. *(Kindly provided by the University of Kansas Museum, Lawrence, Kansas).*

```
QUERY NUMBER 1  S E A R C H  P A R A M E T E R S - - - - - - - - - - - -

EXAMINE CATEGORY 050 OF ALL RECORDS, TURNING ON INDICATOR 3 IF WORD NUMBER 1 IS EQUAL TO          PALMATOLEPSIS

EXAMINE CATEGORY 401 OF ALL RECORDS, TURNING ON INDICATOR 2 IF ANY WORD FOUND IS EQUAL TO         PALMATOLEPSIS

EXAMINE CATEGORY 401 OF ALL RECORDS, TURNING ON INDICATOR 4 IF ANY WORD FOUND IS EQUAL TO         POLYGNATHUS


IF INDICATOR 1 ON OR 2 ON OR 3 ON OR 4 ON
   THEN PERFORM ACTION NUMBER 1.

ACTION NUMBER 1:  OUTPUT CATEGORIES  ALL                                      TO PRINTER

FILE NAME=DEPTPALEO0 INPUT=MASTER    FILE STRUCTURE=NON-HIERARCHICAL
```

Figure 12.9 Example of a SELGEM query. (*Example kindly provided by Office of Computer Services, Smithsonian Institution in Procedures in Computer Science, Vol. 3, June 1975*).

Note that subsets of items resulting from searches through the single categories can be combined by means of Boolean operators. There are options for action to be taken on the subsets: they can be printed out, tallied (that is, only the count is printed out), or replaced so that the retrieval capacity does editing as well. It is possible to ask more than one question at a time. SELGEM has means for retrieving from more than one file in a single search; hierarchical files can be processed by retrieval queries so that data from several subordinate files can be recombined. Thus, SELGEM is adaptable to multilevel data banks. This is an advantage of SELGEM as compared with TAXIR, which only searches a single data table. The linear searches required in SELGEM are less efficient but SELGEM is much better for merging files.

Colombia National Herbarium

Another case where computerized database systems are being used for managing floristic collections is the Colombia National Herbarium (COL) that contained, as of 1982, about 220,000 specimens. Information on about 80 percent of these has been registered in a computerized data bank. A description of the system was published by Forero and Pereira (1976). The purpose of the database is to make the wealth of information available, and at the same time to minimize detrimental handling of specimens. The structure of this data bank is straightforward. Data entries for each specimen form 20 fields that are preplanned to fill the 80 columns of a standard punched card. All of the descriptors (see fig. 12.10 for list), except the names of the geographic locality and of the collector, are coded for entrance into the specimen data bank. Attached to this main data bank is a registry of codes assigned to particular families, genera, and species. Thus, names will be handled in abbreviated form in the computer, but can be retrieved in full when needed.

Output of the program can be obtained in various standard categories and formats. For example, a list can be printed giving the taxonomic names that have been entered with their codes. Or itineraries can be printed arranged by collection number or by date. Complete information on type specimens in a particular group can be put out. Geographic information on different species groups is readily available. Fig. 12.11 is an example of such a printout in which geographic data of different species of the genus *Crotalaria* was the subject of interest.

Museum Documentation System and GOS

In the United Kingdom there has been a major effort over the past decade to standardize museum collection records. The work aims to make the in-

SUMMARY OF SPECIMEN DATA RECORDED AT COL, AND OF THE NUMBER
OF CHARACTERS ASSIGNED TO EACH ITEM

Group (Cryptogams or Phanerogams): 1 digit
Subgroup: 2 digits
    (Algae, Fungi, Mosses, Lichens, Ferns, Gymnosperms, Angiosperms Monocotyledons,
    Angiosperms Dicotyledons)
Creation, Addition or Up-Dating: 1 digit
Family: 4 digits in Form No. 2 (Fig. 2); Max. 22 in Form No. 1 (Fig. 1).
Genus (GEN): 3 digits in Form No. 2; Max. 22 in Form No. 1.
Species (ESP): 3 digits in Form No. 2; Max. 50 in Form No. 1.
Kind of Specimen: 1 digit.
    (Type, Holotype, Cotype, Isotype, Lectotype, Type Var., Paratype, Any other type.
    Regular specimen).
Number of duplicates deposited in the herbarium (No.): 1 digit
    (When more than 9 duplicates are present, the collection is recorded twice).
Herbarium (COL, ICA, MEDEL, INPA, ANT, NY, etc) (H): 1 digit
Country (PAIS): 2 digits
    (American countries were assigned individual codes. Other continents are represented
    by separate codes).
Departamento (DPTO): 2 digits
    (For "Departamentos" and "Municipios" we used the codes already established by
    the DANE-Departamento Administrativo Nacional de Estadística-)
Municipio (MPIO): 3 digits
Habitat (HAB): 2 digits
Geographic Locality (LUG): 2 digits
Name of the Geographic Locality (NOMBRE GEOGRAFICO): 16 digits
Altitude (ALTURA): 3 digits
Abundance (AB): 1 digit
    (Common, Rare, Dominant. Very rare, Very common, Abundant)
Life Form (FV): 1 digit
Uses (U): 1 digit
    (Medicinal, Poisonous, Construction, Ornamental, Food, Spice, Hallucinogenic, Cattle
    Feed, Narcotic)
State of Specimen (E): 1 digit
    (Flower, Fruit, Flower-Fruit, Steril, Fertil, Photo, Drawing)
Collector's Name (COLECTOR): 15 digits
Collection Number (No. DE COLECCION): 7 digits
Collection date (FECHA DE COLECCION): Day: 2 digits; Month: 2 digits; Year: 3 digits

Figure 12.10 Coding of descriptors for Colombia National Herbarium. *(From Forero
and Pereira 1976. Table 1, p. 91).*

formation stored in many national and nonnational museums comparable and
available. The project began, sensibly enough, at the grass-roots level in that
a voluntary association, IRGMA (Information Retrieval Group of the Museum Association), prepared standard record cards suitable for use in all museums. A single multidisciplinary data standard was set and a range of record cards was designed to suit a wide variety of collections such as
archeology, fine arts, geologic specimens, historical artifacts, military artifacts, and natural history specimens. Subscribers to the Museum Documentation System (MDS), which includes computer software as well as standardized documentation, are also free to design their own record cards as

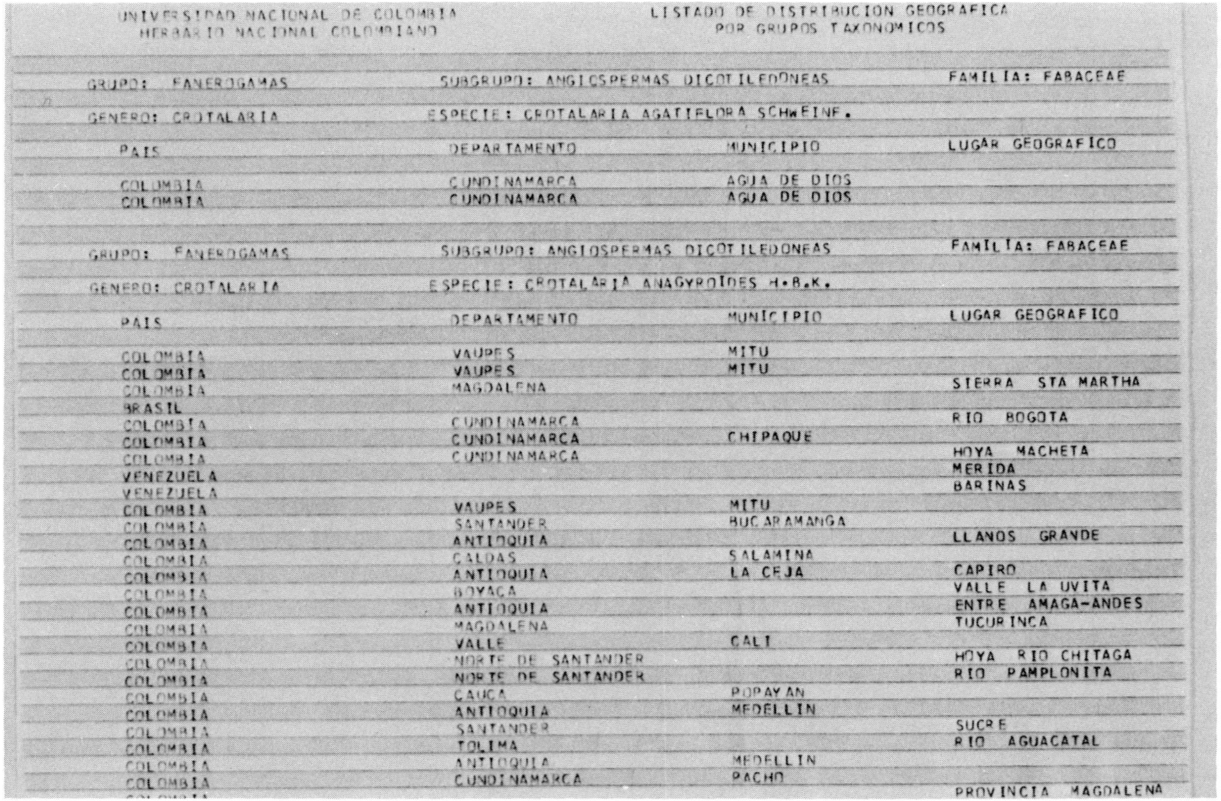

UNIVERSIDAD NACIONAL DE COLOMBIA
HERBARIO NACIONAL COLOMBIANO

LISTADO DE DISTRIBUCION GEOGRAFICA
POR GRUPOS TAXONOMICOS

GRUPO: FANEROGAMAS SUBGRUPO: ANGIOSPERMAS DICOTILEDONEAS FAMILIA: FABACEAE

GENERO: CROTALARIA ESPECIE: CROTALARIA AGATIFLORA SCHWEINF.

PAIS	DEPARTAMENTO	MUNICIPIO	LUGAR GEOGRAFICO
COLOMBIA	CUNDINAMARCA	AGUA DE DIOS	
COLOMBIA	CUNDINAMARCA	AGUA DE DIOS	

GRUPO: FANEROGAMAS SUBGRUPO: ANGIOSPERMAS DICOTILEDONEAS FAMILIA: FABACEAE

GENERO: CROTALARIA ESPECIE: CROTALARIA ANAGYROIDES H.B.K.

PAIS	DEPARTAMENTO	MUNICIPIO	LUGAR GEOGRAFICO
COLOMBIA	VAUPES	MITU	
COLOMBIA	VAUPES	MITU	
COLOMBIA	MAGDALENA		SIERRA STA MARTHA
BRASIL			
COLOMBIA	CUNDINAMARCA		RIO BOGOTA
COLOMBIA	CUNDINAMARCA	CHIPAQUE	
COLOMBIA	CUNDINAMARCA		HOYA MACHETA
VENEZUELA			MERIDA
VENEZUELA			BARINAS
COLOMBIA	VAUPES	MITU	
COLOMBIA	SANTANDER	BUCARAMANGA	
COLOMBIA	ANTIOQUIA		LLANOS GRANDE
COLOMBIA	CALDAS	SALAMINA	
COLOMBIA	ANTIOQUIA	LA CEJA	CAPIRO
COLOMBIA	BOYACA		VALLE LA UVITA
COLOMBIA	ANTIOQUIA		ENTRE AMAGA-ANDES
COLOMBIA	MAGDALENA		TUCURINCA
COLOMBIA	VALLE	CALI	
COLOMBIA	NORTE DE SANTANDER		HOYA RIO CHITAGA
COLOMBIA	NORTE DE SANTANDER		RIO PAMPLONITA
COLOMBIA	CAUCA	POPAYAN	
COLOMBIA	ANTIOQUIA	MEDELLIN	
COLOMBIA	SANTANDER		SUCRE
COLOMBIA	TOLIMA		RIO AGUACATAL
COLOMBIA	ANTIOQUIA	MEDELLIN	
COLOMBIA	CUNDINAMARCA	PACHO	
			PROVINCIA MAGDALENA

Figure 12.11 Sample printout from Colombia National Herbarium. *(Kindly provided by the Colombia National Herbarium via F.A. Bisby).*

Card	File			Institution : identity number				Part	
of				BOLMG : 329.1978					
IDENTIFICA-TION	Simple name	D	Form		Sex	Age	Phase		Number
	Polecat/Ferret		skin		M	adult			
	Classified identification								
	Mammalia & Carnivora & Mustela putorius Linn								
	C	Current ~~Label~~ ~~Other~~	System		Status	D	Identifier : date Berry, K. : 18 Dec 1978			D
	Classified identification								
	C	Current Label Other	System		Status	D	Identifier : date			D
COLLECTION	Place name/detail								
	(ne) Ambleside & Cumbria						vice-county	locality number	
	~~Lat-Long~~ NGR	Other co-ordinates	value & units/accuracy 35.337.031		Altitude Depth	Other position	value & units/accuracy			
	Habitat keyword/detail oak woodland								
	Locality detail A593 nr. Skelwith Bridge								D
	C	Collection method found dead	Collector : date Harding, R. : 12 Dec 1978				Collection number			D
STORE	Store : date Ref. coll - Mammals				Recorder : date Berry, K. : Dec 1978				

NATURAL HISTORY   © IRGMA 1975   1/12/75

ACQUISITION	Acquisition method donated		Acquired from : date Harding, R. (2 The Croft, Threlkeld, Cumbria) : 18 Dec 1978	
C			D Price	Conditions Yes/No D	Valuation : date	D
DESCRIPTION	Condition keyword/detail good		Completeness keyword/detail complete (with skull)	
	Dimension measured weight	value & units/accuracy 1,319 gm	Dimension measured head and body	value & units/accuracy 392 mm
	Part : aspect : description keyword/detail			D
C	tail : dimension : length 141 mm hind foot : dimension : length 66 mm ear : dimension : length 25 mm			
PROCESS	~~Conservation~~ Other process ~~Reproduction~~ preparation	Method/detail : operator : date : detail taxidermy : Berry, K. : 18 Dec 1978		Cross reference: D
DOCUMENTA-TION	L	Class	Author : date : title : journal or publisher : volume : detail	
C				
	L	Class	Author : date : title : journal or publisher : volume : detail	Drawing or photo
C				
NOTES	Notes Found on side of road but no internal bruising or broken bones. Massive haeomorrhage in chest cavity - heart attack. Animal in fine condition but extremely fat.			
C				

Figure 12.12 Examples of MDS record cards. (a) for an animal specimen and (b) for a plant. *(Kindly provided by Museum Documentation Association, Duxford, England).*

long as they conform to the standards. Sample record cards are shown in fig. 12.12. By 1981 the Museum Documentation Association (MDA), which had taken over the work on a full-time basis, had sent out over two million cards for use in 250 museums in the UK, as well as The Netherlands, Tasmania, and Brazil.

A computerized data management system, CGDS (Cambridge Geologic Data System) was developed for geological specimens by J. Cutbill at the Sedgwick Museum. In 1975 this system became the basis for the development of a new program called GOS (a nonacronym name apparently inspired by Robert Burns's name for the goshawk!). Development of GOS was taken over by MDA in 1977, and by the end of that year it was being used to prepare catalogues and indexes from the information on the standard cards. GOS records contain a number of so-called GOS elements, each consisting of a name and associated data (equivalent to what we have called descriptor states). In GOS, such elements can be nested to any extent desired so that the data are stored hierarchically. The length of an element— that is, a data field—is not fixed, but depends on the data held in it.

One of the advantages of standardized data recording is that it greatly facilitates integrated use of data from diverse institutions. MDA is continuing to develop and distribute the GOS software, making it as fully portable as possible. It is usable on both mainframes and 16-bit minicomputers given the correct compiler is available. In addition the MDA computing service is "able to data-prepare and process museum records on behalf of members for whom it can then provide computer-generated catalogues and indexes" (MDA 1980).

Flora Veracruz Project

The Flora Veracruz project in Mexico is a major floristic study of a whole state in eastern Mexico, which lies between the 18,000 foot volcanic peaks east of Mexico City and the Gulf coast. Because of this habitat diversity, there is an unusually rich flora but it is poorly known. The project was one of the pioneers in application of electronic data-processing methods to taxonomic collections. It is an international, collaborative project administered through the Instituto de Nacional Investigaciones sobre Recursos Bioticos (INIREB) and has, as its long-term goal, the production of a complete Flora. Within the project there are several preparatory projects that involve the use of information management systems (Gomez-Pompa and Nevling 1973). One of these deals with the bibliographic database and another with the curatorial tasks involved with recording herbarium and field specimens. In the curatorial database the specimens are the basic items and descriptive data consist of nomenclatural entries, including synonyms, and 32 label data items, including typification and brief comments on morphological features. This information can be made available to taxonomists working on the Flora. In 1981, after more than ten years of work, the data bank for Flora Veracruz contained some 7300 species with a total of 60,000 specimen entries.

The computer program for the curatorial database has the capacity for handling specimen data including printing of labels for any number of duplicates. Once identification is confirmed, proper codes for genus and species are recorded and these, along with label data from the standard field tags, are entered. Locality data are coded according to a grid coordinate system, so that automatic printing of distribution maps like those in fig. 12.2. above, is possible. Once the information is in the file, checklists can be printed and updated easily, and a variety of listings and catalogues can be put out in various report formats. The system allows searching on some fields—those that are codified—but not on fields in text format.

The original database management system at Veracruz was slow at label production because of limitations of the equipment related, in part, to the physical separation of the computer (in Mexico City) and the herbarium (in Xalapa). Recently a Digital VAX minicomputer has been installed at the facility in Xalapa as well as new relational database software (Gomez-Pompa et al. 1984). This permits the capture of data from specimens via terminals and eliminates a step in initial data recording and the need for temporary specimen labels.

The original equipment used at Veracruz also had severe limitations on its capacity for searches and queries. The new system is considerably faster and more flexible. Methods for automatic identification and description that use computerized databases are currently being investigated. A system for identification of the plant families in Veracruz that has already been developed may be used directly on the computer for online identification or identification by comparison. Or the same system may be used indirectly for keying, since it will generate dichotomous keys and center punched polyclaves. In addition a second data bank, based exclusively on vegetative characters, for the identification and description of representative plant species grown in Veracruz is being elaborated. It is hoped that this pilot project will indicate the feasibility of the implementation of a similar program on a large scale.

Floristic Documentation (ESFEDS) and Reference Services (TRF)

In many parts of the world the status of floristic studies is quite different from the situation in Veracruz, and therefore the objectives are different too. In Europe, for example, the present need is not for collecting specimens and making initial records of the plants in the region; data are already available from many studies made over the years. The need now is for correlating the information and making it accessible. The increasing use of computers and other telecommunications devices has provided a fresh impetus for undertaking the task of organizing and communicating this taxonomic data.

A large-scale collation of taxonomic and floristic data on the vascular plants of Europe was begun in 1955, and by 1980 five volumes of *Flora Europaea* had been published. This work is a major database. As such it is the foundation for the European Science Foundation's Taxonomic Documentation System (ESFEDS), which seeks to create a computerized database starting with the information in the Flora and developing records on taxonomic changes. The project was begun in 1982 and is housed at the University of Reading, England. The goal of the first, current phase is to create searchable computer files from the data on nomenclature, geography, and chromosomes. The data included is that in the published work plus more recent data. (New data will simply be added and flagged as such, not incorporated as in a true revision.) In the second phase, databases will be developed for bibliographic data beginning with the core of taxonomic and chromosome references in *Flora Europaea* and extending these with additional references so as to form open-ended bibliographic files. Other databases are being defined in areas such as phytosociology, conservation status of taxa, ecology, weedy plants, phytochemistry, and in biosystematics areas like cytology, genetics, and population variation.

The plan is for telecommunications facilities which will allow direct access for the user. The equipment will enable the user to do online searches for nomenclatural data from remote sources using a Viewdata terminal. Color coding available on these terminals will be used to convey status of a taxon name as, for example, white for accepted name, green for synonyms, and so on. Queries and searches that take more time and involve the main database will be made by using an electronic message service. Here the user request is processed by a database staff member and sent back to the user's terminal at a later time.

An early goal of the ESFEDS project is to produce a catalogue of synonyms of the European plant names. There are other projects that are also concerned with collation of nomenclatural data for taxa in various regions. *Index Kewensis,* the listing of accepted species (and infraspecies) names that was described in chapter 3, is now being computerized. Capture of data was to have been completed by the end of 1983 (Lucas 1984). This will allow direct user access to the names listed with their authorities, native countries, and synonyms. Services such as generic and family lists are planned, either online or using tapes. There is, as a result, the potential for continuous updating and reductions in publishing costs.

The BIOSIS Taxonomic Reference File (BIOSIS-TRF) is another ambitious project whose purpose is to use computers and telecommunications to organize names of taxa and the associated data and references (Dadd and Kelly 1984). Like the ESFEDS European documentation project, the BIOSIS-TRF taxonomic reference project is still in its infancy. Detailed planning is going forward and two pilot projects on bacteria and Vicieae are now underway. The reference file concept has been developed by BIOSIS Inc. (publishers of *Biological Abstracts*) and BIOSIS UK Ltd (their subsidiary). Their original motivation was their need to verify names of biological organisms that are referred to in the scientific literature that they index. Four database components are planned:

1. The Taxonomic Data File will contain names and nomenclatural data. Each name that has appeared in the literature will have its own unique identifying number. These numbers will provide links with alternate names in this database and will form the bases for linkages to associated data in other files.

2. The Hierarchy File will consist of taxonomic classification schemes—multiple schemes in some cases. Each name in the TRF index will be linked to one or more of the taxonomic trees in the Hierachy File.

3. The Related Data File will be a set of taxon by descriptor files that contain descriptive data about each organism such as geographic ranges, common names, and derivative chemicals.

4. The Bibliographic Data File will serve as the link between the entries in TRF and the bibliographic information in BIOSIS' existing bibliographic databases, BIOSIS PREVIEWS and the Zoological Record.

A projected interactive computer-communications system will allow users to search the file, interrogating on the basis of organism name, authority, descriptive data, or a combination of these. Users can copy portions of the file into a work space area in the system to enable them to create versions of the file tailored to their special needs. Such a facility may be useful for producing items such as museum catalogue cards.

A particularly innovative feature proposed for TRF would allow certain approved users to record their scientific findings in special user-comment fields associated with the database, along with appropriate author identification. From time to time, these entries would be evaluated by authorities in the field. A related feature is message exchange between users; that is, a user could record text dealing with new information and indicate as recipients those colleagues whom it might interest. After it is received the message would be erased.

Vicieae Database Project

One of us (FAB) is presently engaged in an experimental monographic database project at the University of Southampton. The purpose of the initial

Vicieae Database Project was to test the feasibility of making a computer database to replace a printed monograph and to experiment with using the computer database to provide improved information services to "user" scientists, who were not necessarily taxonomists. The initial "users" formed an international Study Group of scientists with special interests in the Vicieae. They included phytochemists, a seed expert, a blood transfusion specialist, ethnobotanists, crop biologists and plant breeders.

The tribe Vicieae of the Leguminoseae contains 302 species of vetches and peas. Plants in the Vicieae are important to people as sources of food, fodder, pharmaceuticals, toxic contaminants, weeds, and ornamentals. The project had by late 1981 completed the first phase of getting databases with morphological, chemotaxonomic, and geographical data into operation, although the morphological database was far from complete. A computerized list of accepted species names, and index to synonyms and computer-generated morphological descriptions had been made available.

Much of this work was done using the standard computer programs we have already mentioned—EXIR, CONFOR, GRAPH, and Pankhurst's keymaking and on-line identification programs. R. J. White has written two programs especially for the Project. One is SYNONYMS, which organizes an index of names and synonyms, and the other is EXIRPOST, which reformats and reorganizes output files from EXIR to make them suitable for input to keymaking, cluster analysis, and ordination programs. The flow diagram envisaged for the whole setup is shown in fig. 12.13.

Note that all of the output downstream of EXIR can be produced for any subset of characters and any subset of species. A Portuguese version of the CONFOR character set is on trial for producing descriptions in Portuguese for that country's species in response to a request from a Portuguese user.

The second phase was to experiment with providing novel taxonomic services using the database. Four services provided for members of the Study Group (Bisby 1984d) were a series of cheap, replaceable, published pamphlets; an inquiry service run over a nine-month test period; initial efforts at providing tailor-made products such as catalogues, classifications, and keys to fit requests from particular groups of users; and the provision of parts of the monographic database to other database systems. Members of the Study Group who attended a two-day evaluation meeting at Southampton in September 1983 were generally enthusiastic about the value of both the database and the services provided. They recommended its continued development, particularly emphasizing the value of the services to nontaxonomists (chemists, breeders etc.) and the need for related identification and seed services. However, there were also aspects that it was agreed gave cause for concern: the need for better (coordinated) software, some difficulties in com-

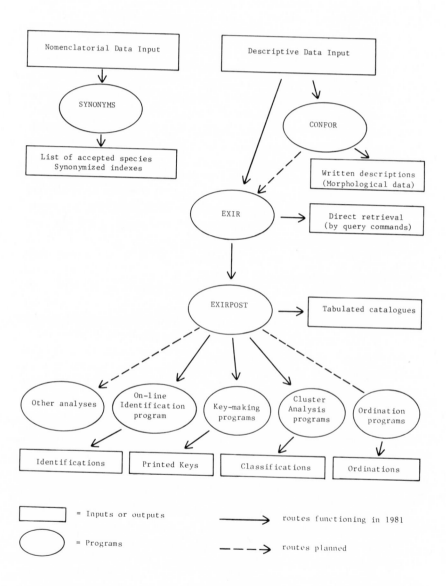

Figure 12.13 Flowchart showing information services provided by the Vicieae Database Project. *(Kindly provided by Vicieae Database Project, Southampton University, Southhampton, England).*

promising between levels of detail suitable for taxonmists and for other scientists, and difficulties in completing the laborious collation of morphological details of necessity fully observed from plant materials.

The experience to date with the Vicieae project has indicated a wide variety of specific interests on the part of users. For example, a plant geneticist has requested information about the presence of anthocyanins in closely related *Lathyrus* species. A chemotaxonomist in Turkey is interested in the distribution of nonprotein amino acids in sections of *Vicia* and *Lathyrus*. In the Vicieae Information Service, unlike some of the planned Documentation and Reference Services, there is no attempt so far to have the user access the database directly; all requests for information are handled by project personel.

Further development of the Vicieae Database is now under way and includes the provision of improved software (see chapter 13 and Allkin 1985), services linked with a seed collection (Bisby et al. 1985), and an experimental use of information for the Vicieae Database as a pilot group for the BIOS TRF.

The Brassicaceae Databank at Notre Dame

A number of rather loosely integrated, largely phytogeographic databases for the family Brassicaceae (which includes cabbage and mustard) are being developed at the University of Notre Dame. The system is referred to as BRASS BAND (meaning BRASSicaceae dataBank At Notre Dame). Although the project is somewhat monographic, in the sense that it deals with a single family and includes some classifications and revisions within the group (for example, in the genus *Cardamine* and the tribe Thelypodiaeae), the major emphasis of the project is on worldwide geographic distributions of taxa in the family.

Components of BRASS BAND include bibliographic, nomenclatural, curatorial, floristic (phytogeographic), and descriptive databases. A major outcome is "A Computer Based Checklist of the Vascular Plants of Indiana" that is based on the multipurpose database entitled INDIANA BRASS. This database evolved from a floristic database (Crovello and Keller 1974), which summarized the distribution of 81 taxa of the Brassicaceae in each of Indiana's 92 counties and which enabled the production of county checklists and maps indicating species diversities. Databases with similar emphasis on distribution of taxa in Brassicaceae that have been established include the NORTH AMERICAN BRASS, EUROPEAN BRASS, and the SOVIET BRASS.

An especially interesting aspect of the Brassicaceae project is the direct use of much of the data in phytogeographic analyses. For example, floristic similarities among 51 regions of the Soviet Union were measured on the basis of the SOVIET BRASS database (Crovello and Miller 1982). The items analyzed were geographic regions (OGU's for operational geographic units) and the characters were distributions for the nine tribes composed of 736 species of Brassicaceae. The geographic units were grouped on the basis of floristic similarities by an ordination based on Principal Components Analysis. A three dimensional plot summarizing the data accounts for 77 percent of the floristic variation.

Analysis of the distribution of points in this plot showed that geographic units tended to form clusters that correlated roughly with superregions of the Soviet Union such as the Arctic, the European area, the Caucasus, East and West Siberia, the Far East, and Central Asia. This suggested that environmental factors may be related to the three dimensions as in an ecological ordination. For example, the second axis may represent a gradient from the dry cold of the Arctic and Central Asian mountains to dry hot regions in parts of Central Asia, Europe, and the Caucasus.

Suggested Readings

General

Allkin, R. and F. A. Bisby 1984. *Databases in Systematics*. London and New York: Systematics Association, Academic Press.
Sarasan, L. and A. M. Neuner 1983. Museum Collections and Computers. Report of the ASC Survey. Lawrence (Kansas): Association of Systematics Collections.

Chapter 13

Goals for the Future

From the start we encouraged readers to think of taxonomy as a general-purpose information system. They were to view the role of models, computers, and databases as a dual one, clarifying the logic and design of the information system as a whole, and providing the practical means of making a modern computer-based system that was better and more efficient than the old. But at both the level of the design and the implementation the task is incomplete; there are weaknesses and gaps.

So in this chapter we run counter to the confidence of earlier chapters by identifying weaknesses in the system as presently envisaged and by outlining, perhaps speculatively, where progress might be made. There is a feeling about that the next few years will see unprecedented changes (Heywood 1984; Bisby 1984 a,b) and as already commented on by Pankhurst (1984), it will be fascinating to see the impact of racing information technology, where devices become obsolete in three to five years, on an ancient information system which measures progress in tens or even hundreds of years.

Gaps in the Computer-Based Information System

In 1968 Rogers and Appan tried to link programs together so that they could manipulate data files on specimens and on species to produce character analyses, cluster analyses, and retrieval of descriptive data for the wild cassavas of the genus *Manihot*. At the time people commented, sometimes unkindly, on the enormous effort that had been put into computerizing each stage. Today what they produced would have been poorly received albeit for different reasons—there were too many restrictions and the software was awkward and unreliable compared with modern systems. But in the taximetrics laboratory at the University of Colorado they were actually taking the first steps toward the same goals that we have today, to produce a coordinated database for taxonomists to manipulate their data, and a coordinated database for providing a taxonomic service to others. Surprisingly, many of the difficulties faced by Rogers and company are only now being overcome in the design of databases.

Coordination

The most important difficulty to overcome is that of coordination between different parts of the software. Fig. 13.1 shows the idealized grand plan: a descriptive database allowing retrieval of data subsets and allowing the subsets to be passed easily and automatically on to various other activities—phenetic cluster analysis or ordination, cladistic analysis, identification aids and description writing. The difficulty is that not only must it be easy to

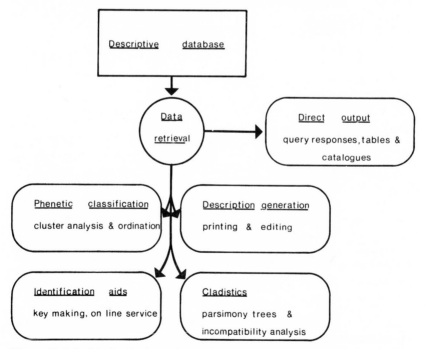

Figure 13.1 An idealized scheme for the use of descriptive information in a variety of taxonomic activities. *(From Bisby 1984a, Fig. 3).*

transfer data from the one program to the next, but when the data are transferred the details that go with the data (descriptor names, types, and states)
must be transferred too. At present many of the best-developed programs for
individual analyses not only require the data in a different format, but also
require a different logical structure or different additional descriptor information. For instance cluster analysis programs require exclusive character-
state data for each item (each item scored for only one character state for
each character) whereas keymaking and description-writing programs accommodate variable entries where each item may be scored for several or a
range of character states for each character.

The need is for a coordinated suite of taxonomic programs that
uses a common data structure and format, or at least a common structure and
format for data transfer. So far the nearest thing to this is the DELTA format
used for data transfer between the KEY keymaking and CONFOR description-writing programs of Dallwitz (as discussed in chapter 12). DELTA format has also been adopted for Pankhurst's identification program suite, and
will probably be available as a transfer format with the BAOBAB database
software being developed at Southampton (Allkin 1985).

Character Handling

It is true that many taxonomic analyses (indeed the majority so far) are performed on a single rectangular data table or flat file of the type described in
chapter 4. But unfortunately, as suggested above, the particular data tables
used in different analyses as part of the same taxonomic study may be different in the form, selection, and expression of characters. Essentially the
same data stored in the database depicted in fig. 13.1 may be required in
different selections, and in different forms (e.g., using sister characters, described in chapter 6) for the activities in each of the remaining boxes in the
figure.

So an additional property of the system should be that the taxonomist can carry out a variety of changes to characters working within the
database. As well as deletions and insertions, a whole range of character
facilities such as alternative versions, dissections, merges, polarity reverses,
and programmed conversions are needed so that a taxonomist can either experiment with characters or do the conversion needed before using the same
data in the different activities.

Dissemination

So far most of the research into taxonomic information systems has concen-
trated on the design and construction of taxonomic databases, and relatively

little work has gone into deciding how to use them in providing improved services to biologists and others. But if the potential of modern information systems is to be realized, a whole new range of styles of dissemination is to be expected. To our knowledge only the Canberra group (Watson, Dallwitz and Macfarlane using microfiche publications) and the Vicieae Database Project (Bisby, White, Allkin and Adey using an inquiry service and pamphlet publications) have experimented with novel services using the databases, and only BIOSIS (with their TRF plans) and ESFEDS (with their Viewdata nomenclatural prototype) have plans for possible international links using telecommunications equipment.

We can envisage people around the world, be they biologists or administrators, gaining access to taxonomic information systems in four ways (Bisby 1985): on-line access, access to distributed copies, off-line enquiry services, and cheap, regularly revised publications.

1) On-line access to a database using a terminal or micro and telecommunications equipment is the method by which the bibliographic databases such as BIOSIS PREVIEWS are used. However it is expensive to use, difficult to organize from developing countries or poorly funded institutions, and requires an enormous commercial demand for the database itself. We know of no taxonomic examples as yet, but this is one of the means being experimented with by the BIOSIS-TRF project.

2) On-line access to a distributed copy of all or part of the database is the method used for the B-I-T-S bibliographic service from BIOSIS, and being considered as another possibility for the BIOSIS-TRF. The advantages in distributing all or part of a database depend on the increasing availability of large micros (and their being easier than telecommunications equipment to obtain and run reliably in developing countries), and the falling cost of storage media such as floppy discs or even video discs that can be copied and sent through the post. What is lost is instant access to the very latest version of a database. Taxonomic databases, however—unlike, say, airline booking systems—are not going to be updated minute by minute; indeed annual revisions may prove more than adequate.

3) Off-line access to an inquiry service is one of the methods by the Vicieae Database Project during an experimental period of providing services to its study group (Bisby 1984d) and planned for its main database by ESFEDS (Heywood et al. 1984). The principal advantages are that (1) users require no equipment; they merely send inquiries by post, telex, or (perhaps in the future) electronic mailbox facilities to an office where a skilled operator queries the database online. (2) The person running the inquiry service can reshape questions to make best advantage of the database. (3) The person running the database can intervene to alert the user to other possible services or, as in the case of the Vicieae Database, to seed collections or other materials that are available (Bisby 1984d; Bisby et al. 1985). Another

feature is that the labor of handling inquiries or making searches is borne by the producers of the database; this may appear to make the service look very much more expensive. But alternatively if the inquiry service is funded by an international agency, it provides a mechanism by which poor or ill-equipped people, often from developing nations, can gain access without costs above those for postage—precisely why this method of access has been chosen for the SEPASAT database on economic plants for the arid tropics: many of their users come from poor, arid regions and part of the project is being funded by the charity OXFAM (Wickens 1984).

Lastly, cheap, updated publications provide limited access to a database but should not be thought of as a substitute for the three methods outlined above. Watson has published microfiche printout in CONFOR description format and published revised versions from time to time. They are relatively quick and cheap to produce from the computer, and can be sent through the post. With a standard microfiche reader one can read through the whole set of generic descriptions from the database. However there is no searching facility: the reader is restricted to a visual scan of the data given in a single physical layout. The same is true of the pamphlets published by the Vicieae Database Project. For instance the synonymized nomenclatural list (Allkin et al. 1983a) or the geographical distribution of *Vicia* by nations (Allkin et al. 1983b) serve a useful purpose by providing a product requested by many users; but once a reader starts searching the pages, say, for names with HB and K as the authority, or for species found in Sicily but not mainland Italy, and not Algeria, they would probably benefit from searching the database itself.

Two Systems or One?

Should taxonomists produce two information systems, an internal technical one containing classificatory data for themselves and the other as an external service providing factual target data for nontaxonomists? Or can one system meet both demands? The question revolves in part on how much the two types of data overlap, and in part on how general or particular are the styles of service needed. By classificatory data we mean the data used by taxonomists to make the classification and provide the nomenclature, identification aids, and descriptions that meet their demands. Target data are wanted by the outside user. Clearly some of the data, such as the micromorphology of pollen surfaces, may be of interest only to taxonomists, while there may be cases where other data, such as gross size, are of interest only to users. Is the overlap—containing, for instance, life form descriptors, geographical distribution, and chromosome numbers—sufficient to make it sensible to make a system for both sets of users? Even where the same data are of interest to taxonomists and users there may be difficulties in how to express the characters. The taxonomist, pigment chemist, and pollination ecologist, while all interested in petal colors of the Boraginaceae, may yet demand quite different expressions of the same data.

       In the past some taxonomic publications have come to grips with this problem. Some recognize the distinction by including a technical diagnosis (with classificatory data) and a general-purpose description for each taxon (for other uses), but others avoid it by the publication serving just the internal or the external audience only. Similar adjustments have yet to be made in taxonomic information systems although the ability of databases to provide different views of the same data may help.

The Future: Brass Knobs and Pure Speculation!

It can do no harm, especially in this new age of burgeoning information technology, to indulge in some wishful thinking and speculation. If unlimited time and money were available what might be done to produce a deluxe taxonomic information system—something with brass knobs on, something which we could be proud of? And why is it that taxonomists expect to be the poor brethren of other scientists, always struggling with poorly funded projects and lagging technology despite the fundamental importance of our work? For in Linnaeus' time the biologists' information system was second to none in efficiency, modernity, and public support. Let us indulge ourselves in three directions—images and graphics, expert systems, and world checklists.

Images and Graphics

We stated in chapter 9 that the two main developments in identification were rather separate—the marked improvement in printed illustrations, and the coming of computerized identification aids. But why cannot the two be combined? The last few years have seen the rapid development of computerized image analysis using both digitized and fourier transform methods for describing shapes, cheap methods for mass storage of images including optical discs, and mass production of graphics reproduction and handling facilities particularly on microcomputers. People are already working on the inclusion and handling of images in databases and in word processing software. To give an example, one of FAB's colleagues, Dr. R. J. White, has developed an image processing system on a CP/M based micro, which was able to record leaf shapes of oak trees using a television camera, score specified leaf shape characters, and submit the scored data table to a canonical variates analysis for a class of students who merely provided leaves collected from nearby trees. The time is already here when we can envisage a coordinated taxonomic information system able to produce illustrations either of organisms themselves, or of characters and character states. It is easy to imagine either technical databases with taxa or character states illustrated by computer stored images, or a variety of popular or educational aids involving the use of images to illustrate, instruct or identify.

Expert Systems

Another developing area of potential use in taxonomy is that of so-called expert systems—computer software that helps in problem-solving if given a set of rules or equipped to learn or deduce a set of rules. Potential areas for applications are in identification, classification, and description generation.

World Checklists

Last, but most important of our extravagant wishes, is that in using databases and telecommunications equipment, the taxonomists and taxonomic institutions around the world may collaborate by contribution to the same world taxonomic information system. This may seem too large a goal for many even to contemplate, but it is undoubtedly present at the back of the minds of people making the most recent suggestions for large-scale international collaborative projects such as the proposal that UNESCO's Man and the Biosphere project might start an international nomenclatural checklist of that project's Biosphere Reserves (Gomez-Pompa, personal communication) or that legume workers collaborate on an International Legume Database (Bisby, 1984c).

A traveler entering Thomas Cook Ltd. in London or American Express in New York, has a booking inquiry dealt with immediately by linked databases that stretch to the corners of the earth—"Frontier from Billings to Great Falls, Montana? Cathay from Hong Kong to Macao? No problem." Why shouldn't an inquirer approaching the New York Botanical Garden or the British Museum receive a similarly efficient access to the vetches of Peru or the bees of Indonesia? The idea of linking up monographic or floristic checklists and eventually full monographic or Floristic databases will surely be an attainable goal in the near future.

References

Abbott, L. A., ed. 1975. *EXIR User's Manual*. Boulder: University of Colorado.

Abbott, L. A. and J. B. Mitton. 1980. Diagnoses generated by numerical taxonomic methods applied to standard blood variables. *Methods of Information in Medicine* 19(4):205–209.

Adams, E. N. 1972. Consensus techniques and the comparison of taxonomic trees. *Syst. Zool.* 21:390–397.

Adams, R. P. 1975. Statistical character weighting and similarity stability. *Brittonia* 27:305–316.

Adey, M. E., R. Allkin, F. A. Bisby, T. D. Macfarlane, and R. J. White. 1984. The Vicieae database: An experimental taxonomic monograph. In Allkin and Bisby, eds. (1984):175–188.

Agnew, A. D. Q. 1968. The interspecific relationship of *Juncus effusus* and *J. conglomeratus* in Britain. *Watsonia* 6:377–388.

Allkin R. 1979. ''The Evaluation and Selection of Plant Characteristics for use in Computer-aided Identification.'' PhD thesis, Polytechnic of Central London.

——. 1984. Handling taxonomic descriptions by computer. Allkin and Bisby, eds. (1984):263–278.

——. 1985. Systematic databases: their use and design. *Science Software Quarterly* (In press).

Allkin, R. and F. A. Bisby, eds. 1984. *Databases in Systematics*. London and New York: Systematics Association, Academic Press.

Allkin, R. and R. J. White 1982. ''Design criteria for a computer program to facilitate the acquisition, storage, retrieval and reformatting of biological descriptions.'' Southampton University (Biology Dept. Internal Report).

Allkin, R., T. D. Macfarlane, R. J. White, F. A. Bisby and M. E. Adey. 1983a. Names and Synonyms of Species and Subspecies in the Vicieae. Issue 2, Vicieae Database Project.

——. 1983b. The Geographical Distribution of *Vicia*. Issue 1, Vicieae Database Project.

Almeida, M. T. and F. A. Bisby. 1984. A simple method for establishing multi-state taxonomic characters from measurement data. *Taxon* 33(3):405–409.

Anderson, E. 1949. *Introgressive Hybridization.* New York: Wiley.
——. 1956. *Genetics in Plant Breeding.* Brookhaven Symposium (BNL 396).
Babac, M. T. and F. A. Bisby. 1984. A Chemotaxonomic Database. In Allkin and Bisby, eds. (1984):209–218.
Baker, E. G. 1914. The African species of *Crotalaria. J. Linn. Soc.* 42:241.
Barkworth, M. E. 1983. *Ptilagrostis* in North America and its relationship to other Stipeae (Gramineae). *Syst. Bot.* 8:395–419.
Barnett, J. A. and R. J. Pankhurst 1974. *A New Key to the Yeasts.* Amsterdam: North Holland.
Barron, D. W. 1984. Current database design—the user's view. In Allkin and Bisby, eds. (1984):35–42.
Baum, B. 1978. *The Genus Tamarix,* Jerusalem: Israel Academy of Sciences and Humanities.
Bentham, G. and J. D. Hooker 1862–1883. *Genera Plantarum.* 3 vols. London: Reeve.
Birch, A. N. E., M. T. Tithecott and F. A. Bisby. 1985. *Vicia johannis* and wild relatives of the faba bean: a taxonomic study. *Econ. Bot.* 39:177–190.
Bisby, F. A. 1970. The evaluation and selection of characters in Angiosperm taxonomy: an example from *Crotalaria. New Phytol.* 69:1149–1160.
——. 1973. The role of taximetrics in Angiosperm taxonomy. 1. Empirical comparisons of methods using *Crotalaria. New Phytol.* 72:699–726.
——. 1981. Genisteae. In Polhill and Raven, eds. (1981):409–425.
——. 1984a. Information services in taxonomy. In Allkin and Bisby, eds. (1984):17–34.
——. 1984b. Automated taxonomic information systems. In Heywood and Moore, eds. *Current Concepts in Plant Taxonomy,* pp. 301–322. London: Academic Press. (Systematics Association Special Volume).
——. 1984c. International legume database. *Bean Bag* 20:3.
——. 1984d. The Vicieae Database Project: Products and services. *Webbia* (In press).
——. 1985. Plant information services for economic plants of arid lands. In G. E. Wickens, ed. *Economic Plants for Arid Lands.* Kew: Royal Botanic Gardens Kew. (In press).
Bisby, F. A. and R. M. Polhill. 1973. The role of taximetrics in Angiosperm taxonomy: 2. Parallel taximetric and orthodox studies in *Crotalaria* L. *New Phytol.* 72:727–742.
Bisby, F. A. and K. W. Nicholls. 1977. Effects of varying character definitions on classification of Genisteae (Leguminosae). *Bot. J. Linn. Soc.* 74:97–121.
Bisby, F. A., J. G. Vaughan, and C. A. Wright, eds. 1980. *Chemosystematics: Principles and Practice.* London and New York: Academic Press.
Bisby, F. A., R. Allkin, B. A. Otto, and M. T. Almeida 1985. Genetic resources of medically important Vicieae (vetches and peas). In M. Mota, ed. *Conservation of Genetic Resources of Aromatic and Medicinal Plants.* Oeiras: Estacao Agronomica Nacional. In press.
Bock, C. E., J. B. Mitton, and L. W. Lepthien. 1978. Winter biogeography of North American Fringillidae (Aves): A numerical analysis. *Syst. Zool.* 27:411–420.
Boratynski, K. and R. G. Davies. 1971. The taxonomic value of male *Cocioidea* (Homoptera) with an evaluation of some numerical techniques. *Biol. J. Linn. Soc.* 3:57–102.

Boltzmann, L. 1898. Vorlesungen Uber Gastheorie, 2. Teil. Leipzig: J. A. Barth.

Boughey, A. S., K. W. Bridges, and A. G. Ikeda 1968. An automated biological identification key. Mus. of Syst. Biol., Univ. of Calif, Irvine. Res. series No. 2.

Boyce, A. J. 1964. The value of some methods of numerical taxonomy with reference to Hominoid classification. In Heywood and McNeill eds. *Phenetic and Phylogenetic Classification.* London: Systematics Association.

Boyce, A. J. 1969. Mapping diversity: A comparative study of some numerical methods. In Coles, ed. (1969): 1–31.

Briggs, D. and S. M. Walters, 1984. *Plant Variation and Evolution.* 2d ed. Cambridge: Cambridge University Press.

Brill, R. C. 1983. "The Taxir Primer, MTS Version." 4th Edition. Ann Arbor: University of Michigan Computing Center.

Brill, R. C. and G. F. Estabrook. 1984. Management of almost flat files in systematic biology using TAXIR. In Allkin and Bisby, eds. (1984): 53–68.

Burnaby, T. P. 1970. On a method for character-weighting a similarity coefficient, employing the concept of information. *Mathematical Geology* 2:25–38.

Burr, E. J. 1970. Cluster sorting with mixed character types 2. Fusion strategies. *Aust. Comput. J.* 2:98–103.

Burt, W. H. 1976. *A Field Guide to the Mammals of America North of Mexico.* Boston: Houghton Mifflin.

Burtt, B. L., I. C. Hedge, and P. F. Stevens. 1970. A taxonomic critique of recent numerical studies in Ericales and *Salvia. Notes R. Bot. Gdn. Edinburgh* 30:141.

Cain, A. J. 1977. Variation in the spire index of some coiled gastropod shells, and its evolutionary significance. *Philos. Trans. R. Soc. Lond. B. Biol. Sci.* 277 (956):377–428.

Cain, A. J. and G. A. Harrison. 1960. Phyletic weighting. *Proc. Zool. Soc. Lond.* 135:1–31.

Camin, J. H. and R. R. Sokal. 1965. A method for deducing branching sequences in phylogeny. *Evolution* 19:311–326.

Cannon, J. F. M. 1964. Infraspecific variation in *Lathyrus nissolia* L. *Watsonia* 6:28–35.

Cavalli-Sforza, L. L. and A. W. F. Edwards. 1967. Phylogenetic analysis: models and estimation procedures. *Evolution* 21:550–570.

Charig, A. J. 1982. Systematics in biology: A fundamental comparison of some major schools of thought. In Joysey and Friday, eds. (1982):363–440.

Chinnery, M. 1973. *A Field Guide to the Insects of Britain and Northern Europe.* London: Collins.

Clapham, A. R., T. G. Tutin, and E. F. Warburg. 1962. *Flora of the British Isles.* 2d ed. Cambridge: Cambridge University Press.

Covell, C. V. Jr. 1984. *A Field Guide to the Moths of Eastern North America.* Boston: Houghton Mifflin.

Clifford, H. T. and W. Stephenson. 1975. *An Introduction to Numerical Classification.* New York: Academic Press.

Coles, A. J., ed. 1969. *Numerical Taxonomy.* New York and London: Academic Press.

Cracraft, J. 1983. The significance of phylogenetic classifications for systematic and evolutionary biology. In J. Felsenstein, ed. *Numerical Taxonomy.* pp. 1–17 Berlin: Springer Verlag.

Cracraft, J. and N. Eldridge. 1979. *Phylogenetic Analysis and Paleontology.* New York: Columbia University Press.

Crisci, J. V. 1980. Evolution in the Subtribe Nassauviinae (Compositae, Mitisieae): A phylogenetic reconstruction. *Taxon* 29:213–224.

Cristofolini, G. and L. F. Chiapella. 1977. Serological systematics of the tribe *Genisteae* (Fabaceae). *Taxon* 26:43–56.

Cronquist, A. 1968. *The Evolution and Classification of Flowering Plants.* London: Nelson.

Crovello. T. J. 1968. Key communality cluster analysis as a taxonomic tool. *Taxon* 17:241–258.

Crovello, T. J. and C. A. Keller. 1974. Uses of computerized floristic data of Indiana for plant geography. *Proc. Indiana Acad. of Sci.* 83:399–406.

Crovello, T. J. and R. D. MacDonald. 1970. Index of EDP-IR projects in systematics. *Taxon* 19:63–76.

Crovello, T. J. and D. C. Miller. 1982. Floristic similarities among 51 regions of the Soviet Union based on the Brassicaceae. *Taxon* 31(3):451–461.

Crovello, T. J., L. A. Hauser and C. A. Keller. 1984. BRASS BAND (The Brassicaceae Data Bank at Notre Dame): An example of database concepts in systematics. In Allkin and Bisby, eds. (1984):219–234.

Dadd, M. J. and M. C. Kelly. 1984. A concept for a machine-readable taxonomic reference file. In Allkin and Bisby, eds. (1984):69–78.

Dahlgren, R. M. T. 1980. A revised system of classification of the angiosperms. *Bot. J. Linn. Soc.* 80:91–124.

Dallwitz, M. J. 1974. A flexible computer program for generating diagnostic keys. *Syst. Zool.* 23:50–57.

——. 1980. A general system for coding taxonomic descriptions. *Taxon* 29:41–46.

——. 1984. Automatic typesetting of computer-generated keys and descriptions. In Allkin and Bisby, eds. (1984):279–290.

Darwin, C. 1859. *On the Origin of Species.* London.

Davis, P. H. and V. H. Heywood. 1963. *Principles of Angiosperm Taxonomy.* Edinburgh and London: Oliver and Boyd.

De Beer, G. R. 1971. *Homology, An Unsolved Problem.* Oxford Biology Readers (No. 11). Oxford: Oxford University Press.

Delany, M. J. and M. J. R. Healy. 1966. Variation in the white-toothed shrews (*Crocidura* spp.) in the British Isles. *Proc. Roy. Soc. B.* 164:63–74.

Du Rietz, G. E. 1930. The fundamental units of biological taxonomy. *Svensk. Bot. Tidskr.* 24:333–428.

Edye, L. A., W. T. Williams, and A. J. Pritchard. 1970. A numerical analysis of variation pattern in Australian introductions of *Glycine wightii. Aust. J. of Agric. Res.* 21:57–69.

Edye, L. A., R. L. Burt, W. T. Williams, R. J. Williams, and B. Grof. 1973. A preliminary agronomic evaluation of *Stylosanthes* species. *Aust. J. Agric. Res.* 24:511–25.

Engler, H. G. A., ed. 1900–1968. *Das Pflanzenreich* (Nos. 1–108). Berlin.

Engler, H. G. A. and K. A. E. Prantl. 1887–1915. *Die natürlichen Pflanzenfamilien.* Leipzig: Engelmann.

Estabrook, G. F. 1966. A mathematical model in graph theory for biological classification. *J. Theoret. Biol.* 12:297–310.

———. 1967. An information theory model for character analysis. *Taxon* 16:86–96.

———. 1968. A general solution in partial orders for the Camin-Sokal model in phylogeny. *J. Theor. Biol.* 21:421–438.

———. 1972. Cladistic methodology: a discussion of the theoretical basis for the induction of evolutionary history. *Ann. Rev. Ecol Syst.* 3:427–456.

———. 1978. Some concepts for the estimation of evolutionary relationships in systematic botany. *Syst. Bot.* 3:146–158.

———. 1979. A Taxir data bank for flowering plant types at the University of Michigan Herbarium. *Taxon* 28:197–204.

———. 1980. The compatibility of occurrence patterns of chemicals in plants. In F. A. Bisby, et al. (1980):379–397.

Estabrook, G. F. and R. C. Brill. 1969. The theory of the TAXIR accessioner. *Math. Biosciences* 5:327–340.

Estabrook, G. F. and C. A. Meacham. 1979. How to determine the compatibility of undirected character state trees. *Mathematical Biosciences* 46:251–256.

Estabrook, G. F. and D. J. Rogers. 1966. A general method of taxonomic description for a computed similarity measure. *BioScience* 16:789–793.

Farris, J. S. 1969. On the cophenetic correlation coefficient. *Syst. Zool.* 18:279–285.

Farris, J. S. 1970. Methods for computing Wagner trees. *Syst. Zool.* 19:83–92.

Farris, J. S., A. G. Kluge, and M. J. Eckardt. 1970. A numerical approach to phylogenetic systematics. *Syst. Zool.* 19:172–189.

Florek, K., J. Lukaszewicz, J. Perkal, H. Steinhaus, and S. Zubrzycki. 1951a. Sur la liason et la division des points d'un ensemble fini. *Colloquium Math.* 2:282–285.

———. 1951b. Taksonomia Wroclawska. *Przegl. Antropol.* 17:193–211.

Forero, E. and F. J. Pereira. 1976. EDP-IR in the national herbarium of Colombia (COL). *Taxon* 25(1):85–94.

Freeston, M. W. 1984. The implementation of databases on small computers. In Allkin and Bisby, eds. (1984):43–52.

Gaffney, E. S. 1979. An Introduction to the logic of phylogeny reconstruction. In Cracraft and Eldridge, eds. (1979):79–112.

Gardiner, G. B., et al. (1979). The salmon, the lungfish and the cow: a reply. *Nature* 277:175–176.

Gibbs Russell, G. E. and P. Gonsalves. 1984. PRECIS—A curatorial and biogeographic system. In Allkin and Bisby, eds. (1984):137–153.

Gleason, H. A. 1968. *The New Britton and Brown Illustrated Flora of the Northeastern United States and Adjacent Canada* (3 volumes). New York and London: Hafner.

Gomez-Pompa, A. and L. I. Nevling. 1973. The use of electronic data processing methods in the Flora of Veracruz program. *Contr. Gray Herb.* 203:49–64.

Gomez-Pompa, A., N. P. Moreno, L. Gama, R. Allkin and V. Sosa. 1984. Flora of Veracruz: Progress and prospects. In Allkin and Bisby, eds. (1984):165–174.

Goodall, D. W. 1966. A new similarity index based on probability. *Biometrics* 22:882–907.

Gower, J. C. 1966. Some distance properties of latent root and vector methods used in multivariate analysis. *Biometrika* 53:325.

——. 1967a. Multivariate analysis and multidimensional geometry. *Statistician* 17:13–28.

——. 1967b. A comparison of some methods of cluster analysis. *Biometrics* 23:623–637.

——. 1971. A general coefficient of similarity and some of its properties. *Biometrics* 27:857–871.

——. 1983. Comparing classifications. In J. Felsenstein, ed. *Numerical Taxonomy*. pp. 137–155. Springer-Verlag, Berlin.

Gower, J. C. and G. J. S. Ross. 1969. Minimum spanning trees and single linkage cluster analysis. *Appl. Statist.* 18:54–64.

Gyllenberg, H. G. 1965. A model for computer identification of micro-organisms. *J. Gen. Microbiol.* 39:401–405.

Hall, A. V. 1970. A computer-based system for forming identification keys. *Taxon* 19:12–18.

——. 1973. The use of a computer-based system of aids for classification. *Contr. Bolus Herb.* 6:1–110.

——. 1975. A system for automatic key-forming. In Pankhurst, ed. (1975):55–63.

Halstead, L. B. 1982. Evolutionary trends and the phylogeny of the Agnatha. In Joysey and Friday, eds. (1982):159–196.

Halstead, L. B., E. I. White and G. T. MacIntyre. 1979. L. B. Halstead and colleagues reply. *Nature* 277:176.

Harris, J. A. and F. A. Bisby. 1980. Classification from chemical data. In Bisby et al., eds. (1980): 305–327.

Hauser, L. A. and T. J. Crovello. 1982. Numerical analysis of generic relationships in Thelypodieae (Brassicaceae), *Syst. Botany* 7(3):249–268.

Hennig. W. 1950. *Grundzuge einer Theorie der phylogenetischen Systematik*. Berlin: Deutscher Zentralverlag.

——. 1966. *Phylogenetic Systematics*. Urbana: University of Illinois Press.

Heywood, V. H. 1984. Electronic data processing in taxonomy and systematics. In Allkin and Bisby, eds. (1984):1–15.

Heywood, V. H., D. M. Moore, L. N. Derrick, K. A. Mitchell and J. van Scheepen. 1984. The European taxonomic, floristic and biosystematic documentation system—an introduction. In Allkin and Bisby, eds. (1984):79–90.

Higgins, L. G. and N. D. Riley. 1970. *A field guide to the butterflies of Britain and Europe*. London: Collins.

Hill, C. R. and P. R. Crane. 1982. Evolutionary aspects of phylogenetic approaches to taxonomic classification. In Joysey and Friday, eds. (1982):269–361.

Hollings, E. and C. A. Stace. 1978. Morphological variation in the *Vicia sativa* L. aggregate. *Watsonia* 12:1–14.

Humphries, C. J. and P. M. Richardson. 1980. Hennig's methods and phytochemistry. In Bisby et al., eds. (1980):353–378.

Hutchinson, J. 1964. *The Genera of Flowering Plants (Angiospermae)*. Oxford: Oxford University Press.

Irwin, H. S. and D. J. Rogers, 1967. Monographic studies in *Cassia* (Leguminosae-Caesalpinioideae). 2. A taximetric study of section Apoucouita. *Memoirs of the N.Y. Bot. Garden* 16:71–118.

Jaccard, P. 1908. Nouvelles recherches sur la distribution florale. *Bull. Soc. Vaud. Sci. Nat.* 44:223–270.

Jardine, N. and J. M. Edmonds. 1974. The use of numerical methods to describe population differentiation. *New Phytol.* 73:1259–1277.

Jardine, N. and R. Sibson. 1968. A model for taxonomy. *Math. Biosci.* 2:465–482.

Jardine, N. and R. Sibson. 1971. *Mathematical Taxonomy*. London: Wiley.

Jeffrey, C. 1977. *Biological Nomenclature*. 2d ed. London: Edward Arnold.

Johnston, B. C. 1980. Computer programs for constructing polyclave keys from data matrices. *Taxon* 29(1):47–52.

Joysey, K. A. and A. E. Friday, eds. 1982. *Problems of Phylogenetic Reconstruction*. London and New York: Academic Press.

Keble-Martin, W. 1965. *The Concise British Flora in Colour*. London: Ebury Press.

Kendrick, B. 1972. Computer graphics in fungal identification. *Can. J. Bot.* 50:2171–2175.

Krauss, H. M. 1973. The Information system design for the Flora North America program. *Brittonia* 25:119–134.

Kruskal, J. B. 1964. Multidimensional scaling by optimizing goodness of fit to a nonmetric hypothesis. *Psychometrika* 29:1–27.

Kupicha, F. K. 1976. The infrageneric structure of *Vicia*. *Notes from the Royal Botanic Garden, Edinburgh* 34:287–326.

Lance, G. N. and W. T. Williams. 1967. A general theory of classificatory sorting strategies. 1. Hierarchical systems. *Comp. J.* 9:373–380.

——. 1968. Note on a new information-statistic classificatory program. *Computer J.* 11:195.

Lawrence, G. H. M. 1951. *Taxonomy of Vascular Plants*. New York: Macmillan.

Legendre, P. 1975. A posteriori weighting of descriptors. *Taxon* 24:603–608.

——. 1978. *EXIR 3.0 Guide de l'Usager*. Montreal: University of Quebec.

Legendre, P. and D. J. Rogers. 1972. Characters and clustering in taxonomy: A synthesis of two taximetric procedures. *Taxon* 21:567–606.

Legendre, P., C. Schreck, and R. J. Behnke. 1972. Taximetric analysis of selected groups of western North American *Salmo* with respect to phylogenetic divergences. *Syst. Zool.* 21:292–307.

LeQuesne, W. J. 1969. A method of selection of characters in numerical taxonomy. *Syst. Zool.* 18:201–205.

——. 1982. Compatibility analysis and its applications. *Zool. J. of Linn. Soc.* 74:267–275.

Linnaeus, C. 1753. (1959) *Carl Linneaeus, Species Plantarum, A facsimile of the first edition 1753*, 2 volumes. London: Ray Society.

——. (1956). *Caroli Linnaei, Systema Naturae, A photographic facsimile of the first volume of the tenth edition (1758)—Regnum Animale.* London: Trustees of the British Museum (Natural History).

Lousley, J. E. and D. H. Kent. 1981. *Docks and Knotweeds of the British Isles.* London: Botanical Society of the British Isles.

Lowry, B., D. Lee, and C. Hebant. 1980. The origin of land plants: a new look at an old problem. *Taxon* 29:183–197.

Lucas, G. Ll. 1984. Databases in systematics: A summing up. In Allkin and Bisby, eds. (1984):321–325.

Macfarlane, T. D. and L. Watson. 1982. The classification of Poaceae subfamily Pooideae, *Taxon* 31:198–203.

McNeill, J. 1975. A generic revision of Portulacaceae tribe Montieae using techniques of numerical taxonomy. *Can. J. Bot.* 53:789–809.

Marriott, F. H. C. 1974. *The Interpretation of Multiple Observations.* London and New York: Academic Press.

Martin, J. 1975. *Computer Data Base Organization.* Englewood Cliffs, N.J.: Prentice Hall.

Mayr, E. 1963. *Animal Species and Evolution.* Cambridge: Harvard University Press.

——. *Principles of Systematic Zoology.* New York: McGraw-Hill.

Mickevich, M. F. 1978. Taxonomic congruence. *Syst. Zool.* 27:143–158.

Milne-Redhead, E. 1961. Miscellaneous notes on African species of *Crotalaria* L. *Kew Bulletin* 15:157–167.

Morse, L. E. 1968. Construction of identification keys by computer (Abstract). *Amer. J. Bot.* 55:737.

——. 1971. Specimen identification and key construction with time-sharing computers. *Taxon* 20:269–282.

——. 1974. Computer programs for specimen-identification, key construction, and description-printing using taxonomic data matrices. *Publs. Mus. Michigan State Univ., Biol. Ser.* 5:1–128.

——. 1975. Recent advances in the theory and practice of biological specimen identification. In Pankhurst, ed. (1975):11–54.

Moss, W. W. 1967. Some analytic and graphic approaches to numerical taxonomy, with an example from the Dermanyssidae (Acari). *Syst. Zool.* 16:177–207.

Moss, W. W., P. C. Peterson and W. T. Atyeo. 1977. A multivariate assessment of phenetic relationships within the feather mite family Eustathiidae (Acari). *Syst. Zool.* 26:386–409.

Nayar, B. K. 1970. A phylogenetic classification of the homosporous ferns. *Taxon* 19:229–236.

Nelson, R. A. 1969. *Handbook of Rocky Mountain Plants.* Tuscon: Dale Stuart King.

Niemala, T. K. and H. G. Gyllenberg. 1975. Simulation of computer-aided self-correcting classification method. In Pankhurst, ed. (1975):137–152.

Orloci, L. 1969a. Information theory models for hierarchic and non-hierarchic classifications. In Coles, ed. (1969):148–164.

——. 1969b. Information analysis of structure in biological collections. *Nature* 223:483–484.

Pankhurst, R. J. 1970. Key generation by computer. *Nature* 227:1269–1270.

——. 1971. Botanical keys generated by computer. *Watsonia* 8:357–368.

——. 1978. *Biological Identification*. London: Arnold.

——. 1984a. On the description of inflorescences. In Allkin and Bisby, eds. (1984):309–320.

——. 1984b. A Review of Herbarium catalogues. In Allkin and Bisby, eds. (1984):155–164.

Pankhurst, R. J., ed. 1975. *Biological Identification with Computers*. Systematics Assoc. Special Vol. no. 7. London: Academic Press.

Pankhurst, R. J. and R. R. Aitchison. 1975. A computer program to construct polyclaves. In Pankhurst, ed. (1975):73–79.

Patterson, C. 1982. Morphological characters and homology. In Joysey and Friday, eds. (1982):21–74.

Payne, R. W. 1975. Genkey: a program for constructing diagnostic keys. In Pankhurst, ed. (1975):65–72.

Payne, R. W., E. Walton, and J. A. Barnett. 1974. A new way of representing diagnostic keys. *J. Gen. Micro.* 83:413–414.

Pielou, E. C. 1969. *An Introduction to Mathematical Ecology*. New York: Wiley.

Polhill, R. M. 1968. Miscellaneous notes on African species of *Crotalaria* L. *Kew Bulletin* 22:169–348.

——. 1981. Taxonomic Part: Supplement to Hutchinson's 'The Genera of Flowering Plants' (1964). In Polhill and Raven, eds. (1981):55–56.

Polhill, R. M. and P. H. Raven, eds. 1981. *Advances in Legume Systematics*. London: Royal Botanic Gardens Kew.

Polunin, O. 1969. *Flowers of Europe: A Field Guide*. London: Oxford University Press.

Polunin, O. and B. E. Smythies. 1973. *Flowers of South-west Europe: A Field Guide*. London: Oxford University Press.

Prance, G. T., D. J. Rogers and F. White. 1969. A Taximetric study of an angiosperm family: generic delimitation in the Chrysobalanaceae. *New Phytol.* 68:1203–1234.

Prentice, H. C. 1979. Numerical analysis of infraspecific variation in European *Silene alba* and *S. dioica* (Caryophyllaceae). *Bot. J. Linn. Soc.,* 78:181–212.

Prim, R. C. 1957. Shortest connection networks and some generalizations. *Bell System Tech. J.* 36:1389–1401.

Rempe, U. and E. Weber. 1972. An illustration of the principal ideas of MANOVA. *Biometrics* 28:235–238.

Rogers, D. J. 1963. Taximetrics: new name, old concept. *Brittonia* 15(4):285–290.

Rogers, D. J. and S. G. Appan. 1973. *Manihot, Manihotoides. (Euphorbiaceae) (Flora Neotropica, Monograph No. 13.)* New York: Haffner Press.

Rogers, D. J. and H. S. Fleming. 1973. A Monograph of *Manihot esculenta* with an explanation of the taximetrics methods used. *Econ. Bot.* 27:1–113.

Rogers, D. J. and C. F. Rogers. 1978. Systematics of North American Grape Species. *American Journal of Enology and Viticulture* 29:73–78.

Rogers, D. J. and T. T. Tanimoto. 1960. A computer program for classifying plants. *Science* 132:1115–1118.

Rohlf, F. J. 1962. "A numerical taxonomic study of the genus *Aedes* (Diptera: Culicidae) with emphasis on the congruence of larval and adult classifications." PhD thesis, University of Kansas.

——. 1968. Sterograms in NT. *Syst. Zool.* 17:246–255.

——. 1972. An empirical comparison of three ordination techniques in numerical taxonomy. *Syst. Zool.* 21:271–280.

——. 1974. Methods of comparing classifications. *Ann. Rev. Ecol. and Syst.* 5:101–113.

——. 1982. Consensus indices for comparing classification. *Math. Biosci.* 59:131–144.

Rosen, R. 1959. The DNA-protein coding problem. *Bull. Math. Biophysics* 21:71.

Ross, G. J. S. 1969. Minimum spanning tree. *Appl. Statist.* 18:103–104.

Ross-Craig, S. 1948–1974. *Drawings of British Plants.* London: Bell & Sons.

Saldanha, C. J. and C. K. Rao. 1975. *A Punched Card Key to the Dicot Families of South India.* Bangalore: Center for Taxonomic Studies, St. Joseph's College.

Sarasan, L. and A. M. Neuner. 1983. *Museum Collections and Computers. Report of an ASC Survey.* Lawrence (Kansas): Association of Systematics Collections.

Schauer T. 1982. *A Field Guide to the Wild Flowers of Britain and Europe.* London: Collins.

Schnell, G. D. 1970. A phenetic study of the suborder Lari (Aves). 1. Methods and results of principal components analyses. *Syst. Zool.* 19:35.

Schultz, L. 1983. Computerized literature file. *Medical Electronics* 14:72–78.

Shannon, C. E. 1948. A mathematical theory of communications. *Bell Syst. Tech. J.* 27:379–423, 623–656.

Shepard, R. N. 1980. Multidimensional scaling, tree-fitting and clustering. *Science* 210:390–398.

Shetler, S. G. 1975. A generalized descriptive data bank as a basis for computer-assisted identification. In Pankhurst, ed. (1975):197–236.

Simpson, G. G. 1961. *Principles of Animal Taxonomy.* New York: Columbia University Press.

Small, J. K. 1933. *Manual of the Southeastern Flora.* Chapel Hill: University of North Carolina Press.

Sneath, P. H. A. and R. R. Sokal. 1973. *Numerical Taxonomy.* San Francisco: W. H. Freeman.

Sokal, R. R. and P. H. A. Sneath. 1963. *Principles of Numerical Taxonomy.* San Francisco: W. H. Freeman.

Sokal, R. R. and C. D. Michener. 1958. A statistical method for evaluating systematic relationships. *Kans. Univ. Sci. Bull.* 38:1409.

——. 1967. The effects of different numerical techniques on the phenetic classification of bees of the *Hoplitis* complex (Megachilidae). *Proc. Linn. Soc. Lond.* 178:59–74.

Sokal, R. R. and F. J. Rohlf. 1962. Cophenetic comparisons of dendrograms. *Taxon* 11:33–40.

——. 1969. *Biometry.* San Francisco: W. H. Freeman.

——. 1980. An experiment in taxonomic judgement. *Syst. Bot.* 5:341–365.

Sosa, V. 1978. *Flora de Veracruz, Fascículo 2. Cornaceae.* Xalapa: Instituto de Investicaciones sobre Recursos Bióticos.

Stacc, C. A. 1980. *Contemporary Biology: Plant taxonomy and biosystematics.* Baltimore: University Park Press.

Stearn, W. T. 1971. A survey of the tropical genera *Oplonia* and *Psilanthele* (Acanthaceae). *Bull. of the Brit. Mus. Nat. Hist. Bot.* 4:261–323.

Sweet, H. C. and J. E. Poppleton. 1977. An EDP technique designed for the study of the local flora. *Taxon* 26(2/3):181–190.

Symonds, G. W. D. 1963. *The Shrub Identification Book.* New York: Morrow.

Theophrastus 1916. *Enquiry into Plants.* Translation by A. Hort (2 vols.). London and New York: Loeb Classical Library.

Thorne, R. F. 1976. A phylogenetic classification of the Angiospermae. *Evolutionary Biology* 9:35–106.

Thorpe, R. S. 1980. Microevolution and taxonomy of European reptiles with particular reference to the grass snake *Natrix natrix* and the wall lizards *Podarcis sicula* and *P. melisellensis. Biol. J. Linn. Soc.* 14:215–234.

Throckmorton, L. H. 1968. Concordance and discordance of characters in *Drosophila* classification. *Syst. Zool* 17:355.

Tukey, J. W. 1977. *Exploratory Data Analysis.* Reading, Mass.: Addison-Wesley.

Turrill, W. B. 1935. The investigation of plant species. *Proc. Linn. Soc. Lond.* 147:104–105.

Tutin, T. G., V. H. Heywood, N. A. Burges, D. M. Moore, D. H. Valentine, S. M. Walters and D. A. Webb, eds. 1964–1980. *Flora Europaea.* Cambridge: Cambridge University Press.

Van Dam, A. and D. E. Rice. 1971. On line text editing: A survey. *Comput. Surv.* 3:93–114.

Wagner, W. H. Jr. 1963. Biosystematics and taxonomic categories in lower vascular plants. *Regnum Vegetabile* 27:63–71.

Walters, S. M. 1961. The shaping of Angiosperm taxonomy. *New Phytol.* 60:74–84.

Watson, L. 1967. A mixed-data numerical approach to angiosperm taxonomy: the classification of Ericales. *Proc. Linn. Soc. Lond.* 178:1–25.

Watson, L. and Dallwitz, M. J. 1980. *Australian Grass Genera—Anatomy, Morphology and Keys.* Canberra: The Australian National University, Research School of Biological Sciences.

Watson, L. and P. Milne. 1972. A flexible system for automatic generation of special purpose dichotomous keys and its application to Australian grass genera. *Aust. J. of Bot.* 20:331–352.

Watson, L., W. T. Williams and G. N. Lance. 1966. Angiosperm taxonomy: a comparative study of some novel numerical techniques. *J. Linn. Soc.* 59:491.

——. 1967. A mixed data approach to Angiosperm taxonomy: The classification of Ericales. *Proc. Linn. Soc. Lond.* 178:25–35.

Weber, W. A. 1975. Polyclave key to the species of *Astragalus* in Colorado. (Punched card key.) Boulder: University of Colorado Museum.

Webster, T. 1969. Developments in the description of potato varieties. Part 1—Foliage. *J. Natn. Inst. Agric. Bot.* 11:455–475.

——. 1970. Developments in the description of potato varieties. Part 2: Inflorescences and tubers. *J. Natn. Inst. Agric. Bot.* 12:17–45.

Whiffin, T. 1982. Numerical analysis of volatile oil data in systematic studies of Australian rainforest trees. *Taxon* 31:204–210.

Wickens, G. E. 1984. Plants for man—the Kew Data Bank of Economic Plants. *International Relations* 8:73–80.

Wiley, E. O. 1981. *Phylogenetics: The Theory and Practice of Phylogenetic Systematics.* New York and Chichester: Wiley.

——. 1978. The evolutionary species concept reconsidered. *Syst. Zool.* 27:17–26.

Williams, W. T., H. T. Clifford and G. N. Lance. 1971. Group-size dependence: a rationale for choice between numerical classifications. *Comp. J.* 14:157–162.

Williams, W. T. and J. M. Lambert. 1959. Multivariate methods in plant ecology. 1. Association analysis in plant communities. *J. Ecol.* 47:83–101.

——. 1960. Multivariate methods in plant ecology. 2. The use of an electronic digital computer for association analysis. *J. Ecol.* 48:689–710.

Wilson, E. O. 1965. A consistency test for phylogenies based on contemporaneous species. *Syst. Zool.* 14:214–220.

Wirth, M., G. F. Estabrook and D. J. Rogers. 1966. A graph theory model for systematic biology, with an example for the Oncidiinae (Orchidaceae). *Syst. Zool.* 15:59–69.

Woodell S. R. J. 1965. Natural hybridization between the cowslip (*Primula veris* L.) and the primrose (*P. vulgaris* Huds.) in Britain. *Watsonia* 6(3):190–202.

Index

Note: **Bold face** *numbers refer to figures*

Algorithms: definition of, 73; agglomerative polythetic, 114; agglomerative polythetic-optimization, 115; monothetic devisive, 115
Angular distance measure, 79, **5.3**
Annotation, 54
Authority, 30

Binary data, 64
Binomial, 28
BioScience Information Service (BIOSIS): *Biological Abstracts,* 38, 287; BIOSIS PREVIEWS, 284, 317; BIOSIS-TRF, 308, 317; B-I-T-S, 284, 317; *Zoological Record,* 39
Brassicaceae Databank of Notre Dame, 312

Canonical variates analysis; *see* MANOVA
Central point cluster analysis: median cluster analysis, 159, **7.12, 7.13;** true centroid, 159, **7.14**
Character analysis: qualitative data, 104–108, 132–137; quantitative data, 137–140
Character dependencies, 62, 289, **12.4**
Character states: balanced vs. rare, 125; frequency distributions of, 68, 127–129, **4.3, 6.1, 6.2, 6.3;** in relation to information, 101 ff., 124 ff., **5.11, 5.12, Table 6.1**
Characters: classificatory value of, 121; definition of, 42; incompatible, 252; in relation to descriptors, 59; selection of, 140–142; types of, 122 ff.

CHARANAL, 132–137
Chaining, 180, **7.26(a)**
Circle criterion, for linkage diagrams, 204, **8.13**
Cladogram or cladistic tree, 229, **10.3;** and corresponding Venn diagrams, **10.17;** in relation to various trees, **10.4, 10.5**
Cladism, 243–251
Classical cladists, 250
Classification: artificial, 22; categories in, 21; definition of hierarchical 13; natural, 11, 22; special purpose, 17; standard methods of, 48
Cliques, 252, **10.22**
Cluster analysis: agglomerative two-step, 145, **7.1;** group-size dependence of, 183; optimization, 165 ff.; stability of different methods of, 181
Cluster shape, 178, **7.22**
Coding (of descriptions), 291–293, **12.5, 12.6**
Coding (of states), 261, 301, **Table 4.2, 12.10**
Colombia National Herbarium, 301, **12.10, 12.11**
Compatibility, 251–254
Components, principal, 83
Computer plotting: of species distribution maps, **12.2;** stereoscopic, **8.8;** 2-D, 190, **8.2;** 3-D perspective, 197, **8.7**
Computers: use in taxonomy, 2, 7 ff.; micros, minis, and mainframes, 257–259
CONFOR, 292, 318, **12.6(a)**
Contingency table, 148, **Table 7.1**
Consensus trees, 183, **7.28**

Convergence, 230; parallel change, 230; reversal, 230; true convergence or mistaken homology, 231
Cophenetic correlation, 181
Correlation, 80, 81; of classificatory characters, 47, 129 ff., **6.4;** detection of in quantitative characters, 137–140; as measure of distance, 80; measure of in qualitative characters, 134; statistical vs. taxonomic, 81, **5.4**

DFA, 87 ff., **7.19;** frequency diagrams for, **5.7(c)**
Data collection, 260; forms for , **11.1(a), (b)**
Data matrix, 61
Data storage: editing and updating, 268 ff.; input (TAXIR example), 266–268
Data structure: nontabular, 69; tabular, 59 ff.
Data types, 63 ff., **4.1**
Database management: choice of systems, 281 ff.; computerized (principles of), 261–275; hierarchical, 275; network, 277; relational, 278–280
Databases, 8, 259; bibliographic, 284–287, **12.3;** biogeographic, 284, **12.1, 12.2;** curatorial, 284; descriptive, 288–293; major examples of, 293–313; nomenclatural, 288
DELTA, 291 ff., **12.6(a)**
Dendrograms, 98, 197–202, **5.10, 7.25, 8.9;** representations of clustering results, **7.3–7.8;** truncated, 87, 201, **5.7(b), 8.10**
Descriptions: coded, 291, **12.5, 12.6;** popular, 24, **2.2;** technical, 24, **2.1**
Descriptors (variables), names and states of, 59
Dichotomies (in keys), 31
Discriminant Function Analysis, *see* DFA
Distance measures: angular, 79; Euclidean, 76, 147; Manhattan block, 78, 147

Eigenvalues, 85
Eigenvectors, 85
Epithet, 28
Equivalence classes, 96
ESFEDS, 308, 317
Euclidean distance, 76, **5.1;** calculation of, **5.2**

Field guides, 212 ff.
Fields, 259
Files, 259; flat, 259, 264; merging, 275; multiple, 275

Flora Veracruz Project, 306
Floras, Faunas, 20, 24, 34, **2.1**
Floristic documentation, *see* ESFEDS
Form, 29
Format, input, 261
Frequency distribution, 68

Generalized distance (Mahalanobis D^2), 90, 171
GOS, 306
Graph: directed, 97; nodes, edges of, 94, **5.8;** undirected, 97, **5.8**
GRAPH, 296
Gray Card Index, Harvard University, 38
Group-average cluster analysis, 157 ff., **7.10**

H, *see* information measures
Histogram, 68; bimodal, **4.3, 6.1;** unimodal, **6.2**
Homology, 42, 60
Homoplasy, 231, 248
Hyperspace model, 75

Identification, 211 ff.; by computer, 222; elimination methods (*see* keys, polyclaves); matching methods, 223, **9.6**
Identification aids, *see* keys, polyclaves
Index Kewensis, 38, 308
Information gain, 111, **5.15**
Information measures: calculation of, **5.12;** calculation of maximal, **5.11;** for characters, H_a, 101; for character comparisons, 104–108, **5.13(a), (b), and (c);** of diversity, 109, **5.14**
Information retrieval: in flat files (TAXIR example), 270–274, **11.5(a–c);** in multiple files (GIS example), 275–277, **11.6;** in a relational database, 278–280, **11.7**
Information services, 19, 317, **1.1**
Internal link, in linkage groups, 154, 204, **7.7**
Interval data (linear or nonlinear), 63

Keys (dichotomous or single entry), 30 ff., 213; balanced and unbalanced, **9.1;** bracketed, 31; computer-produced, 214–217, 225 ff.; condensed, 215, **9.2;** indented, 32

Leads (in keys), 31
Linkage diagrams: drawing, 202–204, **8.12;** examples, **7.2–7.9;** in relation to dendrogram representations, 205, **8.14**

Linkage groups (*see also* subgraphs), 95
Loadings, on principal components, 85

MANOVA, 171–177, **7.20, 7.21**
MDA, *see* Museum Documentation System
MDS, 91, 170, **7.18**
MST, *see* minimum spanning tree
Manhattan or city block distance, 78, 147; applied to evolutionary distances, **10.13**
Metric data, 63
Metric distance measures, 78
Minkowski metrics, 78
Minimum spanning tree, 97, 207, **5.9**; combined with scatter diagram, 208, **8.15, 8.16**; two styles of drawing, **8.15**
Models, 73; geometric, 74–92; graph and set theoretic, 92–99; information theory, 99–112
Monographic revisions, 13, 20, 34, 35
Monographs, 35–37
Monothetic divisive clustering method, 165
Monothetic groups, 93
Multidimensional geometric model, 75
Multiple access, 271
Multistate data: binary, asymmetric and symmetric, 64; ordered, 64, **Table 4.2;** ordered, linear or nonlinear, 65; simple, 64, **Table 4.2**
Multivariate data tables, analysis of, 190–193
Museum Documentation System, 301–306, **12.12**

Names: common, 14; for families, 27, 29; for genera, 27; for species, 28; for subspecies, varieties, 28, 29
Names, publication of, 52
Narrative relationship diagrams: balloon, **10.7;** line, **10.9;** tree, **10.8, 10,10**
Nomenclature, 27–30, 52; effective publication, 52; rules for, 27, 30; valid publication, 52
Nonmetric multidimensional scaling, *see* MDS
Notation system, 60

Object (or item) definition, 59
Ordinal data, 63
Ordination, 145, 166; ecological, 19; plotting in reduced space, 86
Orthogonal axes, 76

PCA, 83 ff., **5.5, 5.6, 5.7**
PCO, 90, 167–170, **7.17**
Pairwise data, 69
Parallelisms, 230, 246, 251 ff., **10.21**
Parsimony, 237–242
Partitioning, 68
Pattern cladists, 250
Phenetic and phyletic patterns, 22, 229
Phylogenetic reconstruction: cladism, 242–251; compatibility methods, 251–254; narrative methods, 233–237; parsimony methods, 237–242; relation to taxonomy, 17
Physical models for 3-D data, 193, **8.5**
Polyclaves (multiple-entry keys), 217; edge-punched card key, **9.3;** computer produced, 223, **9.5;** punched card key, 221, **9.4**
Polythetic groups, 93
Principal Components Analysis, *see* PCA
Principal Coordinates Analysis, *see* PCO
Probability: conditional, 104; joint, 108
Product moment correlation coefficient, r, 80, 104, 148

Q analysis, 109
Qualitative data, 64
Quantitative data: discontinuous, 63; metric, 63
Quasimetrics, 78
Querying, 269

r, *see* product-moment correlation coefficient
R analysis, 109
Ranks, 21; determination of, 49, 51
Records, 259
Reference services, *see* TRF
Refinement, 49
Relation: definition of, 95; equivalence, 96
Relational database management, 278–280
Relations (in relational databases), 278
Resemblance measures, *see* distance, similarity measures
Reversals (in central point clustering), 163, **7.15**
Reversals (in evolutionary descent), 230, 246, 251 ff., **10.21**

Scaling, 68
Scatter diagrams, 188 ff.; examples of, **5.7(a), 6.6, 6.7, 6.8, 8.1–8.3;** pictorialized, 196, **8.6;** use of for detecting correlations in quantitative data, 137–140

Searching, *see* information retrieval
SELGEM, 298, **12.8, 12.9**
SEPASAT, 318
Semimetrics, 78
Similarity measures, definition and properties, 94, 95; Gower's mixed data coefficient, 67, 150; Rogers and Estabrook's mixed data coefficient, 150; nonmetric coefficient, 149; simple matching coefficient, 149
Single-linkage: cluster analysis, 150–156, **7.3–7.7;** recommendation for use above species level, 186; rule, 114
Space contraction, 179, **7.23**
Space dilation, 179, **7.24**
Species: concepts (biological, morphological, evolutionary), 50; endangered, 15
Standardizing, 68
Steiner minimum spanning tree, 240
Steiner points, 240, **10.14**
Subgraph: definition of, 96, **7.2;** maximal connected, **5.8;** relation to linkage diagrams, 151
Sums of squares agglomerative clustering, 165
Synapomorphies, 243–249, **10.18–10.20;** apomorphous (derived) state, 243, **10.17;** plesiomorphous (primitive) state, 243, **10.17**
Synonyms, 30

Taxa (taxon), 21
TAXIR: description of, 264–274; examples of use, 296, **12.3, 12.7**
Taxonomic information system, 13, 283; access to, 12; elements of, 20; gaps in, 314–318; Linnaeus', 11; uses of, 14 ff.
Tree: directed and undirected, 97, 237, **5.10, 10.12;** rooted, 97
TRF (Taxonomic Reference File), 309
Type specimens, 54
Typification, 54

Ultrametrics, 99
University of Colorado: herbarium, 296; taximetrics laboratory, 315
University of Kansas, Museum of Natural History, 298
University of Michigan, herbarium, 296

Variation (within taxon), 62
Venn diagrams, 133, 243, **6.5, 10.17, 10.18, 10.19**
Vicieae Database Project, 309–312, 317, **12.13**

Wagner trees, 240
Wagner network, 239–240, **10.15**
Weighting (of characters): a posteriori, 45, 121, 141; a priori, 81, 121, 140
WISS, 239